INSOMNIA
A Guide for Medical Practitioners

INSOMNIA

A Guide for Medical Practitioners

A. N. Nicholson

OBE, DSc, MB, ChB, PhD, FRCPath, MFOM
Consultant in Aviation Medicine
Royal Air Force Institute of Aviation Medicine, Farnborough

J. Marks

MA, MD, FRCP, FRCPath, MRCPsych
Director of Medical Studies
Girton College, Cambridge

1983 **MTP PRESS LIMITED**
a member of the KLUWER ACADEMIC PUBLISHERS GROUP
BOSTON / THE HAGUE / DORDRECHT / LANCASTER

British Library Cataloguing in Publication Data
Nicholson, A. N.
  Insomnia
  1. Insomnia
  I. Title   II. Marks, J.
  616.8'49   RC548

  ISBN 978-94-010-9430-6   ISBN 978-94-010-9428-3 (eBook)
  DOI 10.1007/978-94-010-9428-3

Library of Congress Cataloging in Publication Data
Nicholson, A. N.
  Insomnia, a guide for medical practitioners

  Includes index.
  1. Insomnia.   2. Sleep.   I. Marks, John 1924–
  II. Title. [DNLM: 1. Insomnia. 2. Insomnia—therapy. 3. Sleep.
  WM 188 N624I]
  RC548.N53   1983   616.8'49   83–22233
  ISBN 978-94-010-9430-6

Typeset by Northumberland Press Ltd, Gateshead

Contents

Acknowledgements

The authors are indebted to many of their colleagues who have unconsciously contributed to this book. In particular to Barbara Stone, Douwe Breimer, William Dement, Peter Hauri, Thomas Roth and to Mrs Adrienne Bagwell who prepared the manuscript.

PART I

Sleep – Normal and Abnormal

Chapter 1

Sleep

Introduction

For centuries the enigma of sleep has held a considerable fascination for man. This is not surprising for we spend about one-third of our lives asleep. It is a form of behaviour that is almost universal throughout the animal kingdom and so must, presumably, have some relevance to our welfare. Yet the significance of sleep is far from clear. Indeed, it is still difficult to provide a satisfactory definition.

Sleep can be regarded as an inate form of behaviour which removes the individual from the surroundings in which it is necessary to maintain activity essential for survival. Hence the concept that sleep has a protective function. It has also been suggested that sleep conserves energy and allows restoration of the normal wake function, though if we look carefully at the restorative function of sleep, there is little evidence of this in physical terms.

For example, during sleep deprivation a variety of blood, urine and cardiovascular measurements show no changes. Nor does sleep deprivation lead to any change in the capacity for physical work of the type that might be expected if restoration of muscle activity were a major effect. It is true that during sleep the plasma level of growth hormone rises, and that this could be regarded as anabolic and hence restorative, but studies on protein balance during sleep show a net decrease in tissue protein.

In fact, on the physical side, the concept that sleep is entirely beneficial may be viewed with scepticism. People with very long sleep periods have been shown to be more prone to cerebrovascular accidents than those with short sleep periods. Moreover, sleep apnoea, dealt with later in this book, presents a significant problem.

However, within the field of mental activity sleep is beneficial, as deprivation leads to irritability, speech slurring and minor visual hallucinations. Personal relationships may also deteriorate. Complex tasks which require active attention show little change, but mundane tasks show deterioration. This suggests that adequate incentives will overcome sleep performance decrements. Nevertheless, mental fatigue is more likely to necessitate sleep than physical fatigue which will respond adequately to rest.

Mental fatigue is most liable to occur as a result of activities which demand sustained attention. Nevertheless, we are all well aware that tiredness is most pronounced after a day of frustration rather than after a day of successful and fulfilling activity. Thus the behaviour of sleep and wakefulness depends in large measure on emotional content.

But whatever its function and need, sleep is a pleasant and relaxing phenomenon enjoyed and desired by most. Sleep satisfaction is an important desire, and the management of sleep disorders might be effected more easily and rationally with less recourse to drug therapy if the accent could be placed on sleep satisfaction. Since there is no exact requirement for sleep it follows that the complaint of insomnia is essentially one of 'sleep dissatisfaction'. Insomnia is not a disease but a symptom, and, like all symptoms, its cause must be sought before any attempt is made to correct the situation.

Lack of sleep is subjectively unpleasant, and no less than one in three adults experience some difficulty in sleeping at some time each year – delayed onset, early awakening, insufficient or broken sleep. Of those who experience sleep problems about 60% have difficulty getting off to sleep without the aid of sleeping pills, about 20% complain of waking during the early hours of the morning and difficulty in getting to sleep again, and the remainder experience both these problems. With the strong need for untroubled sleep demanded by human nature, the high level of sleep dissatisfaction that exists in the community and the initial good results that can be achieved with hypnotics, it is scarcely surprising that hypnotics account for some 5% of all prescribed drugs.

On the basis of a high level of sleep dissatisfaction treated extensively with hypnotics, the aims of this book are to:

(1) Consider our current understanding about the nature of sleep,
(2) Describe the common and less common sleep disorders,
(3) Consider when treatment for sleep disorders is necessary,
(4) Describe how they may best be treated, and above all
(5) Consider how hypnotics should be used in a rational manner.

To achieve these aims we have considered the needs of practitioners in family medicine, where the vast majority of patients who complain of sleeping difficulties have transient insomnia – the result of changed environment, a temporary situational problem or an acute emotional crisis. With careful management these patients can be rapidly cured and chronic insomnia avoided, and our overall aim has been to explain how the symptom of insomnia can be best managed.

The Physiological Mechanisms of Sleep and Wakefulness

Consciousness of our surroundings during wakefulness depends on sensory messages reaching the cerebral cortex and being adequately interpreted. Indeed, the state of wakefulness does not depend directly on the activity of the cells of the cerebral cortex, but on the influence exerted by three clusters of brain stem cells. One of these clusters is known as the reticular activating system. As long as activity in this system is high it will send impulses widely to the cortex via thalamic nuclei and maintain a state of wakefulness. Although there appears to be an inherent rhythm within the cells of the reticular activating system – periods of activity alternate with periods of quiescence – its activity can be controlled by other parts of the nervous system.

A wakeful state is required when sensory messages need to be interpreted. The reticular formation is itself stimulated by nerve paths which are activated at the same time as the main nerve fibres, and the influences that excite the reticular activating system include:

(1) Sensory stimuli – especially pain,
(2) Voluntary or willed impulses to stay awake,
(3) Unusual visual, auditory and olfactory stimuli, and
(4) The basic drives of hunger, thirst and mating.

When the level of activity in this system is low, then relaxed wakefulness and possibly even drowsy sleep may supervene. But a deeper stage of sleep requires not only reduced activity in the reticular activating system, but also active involvement of a sleep system which is believed to involve the serotonergic nuclei of the raphe system of the pons and the medial forebrain area.

Factors which reduce activity in the reticular activating system, but encourage sleep include:

(1) Monotonously repeated auditory and visual stimuli – hence the soporific effect of some lecturers and preachers.
(2) Warmth,
(3) Voluntary influences from higher centres. Some of these are of a conditioned nature – hence a ritual undertaken by many people to encourage sleep,
(4) Impulses from internal organs – as with a full stomach or the satisfied urge to mate.

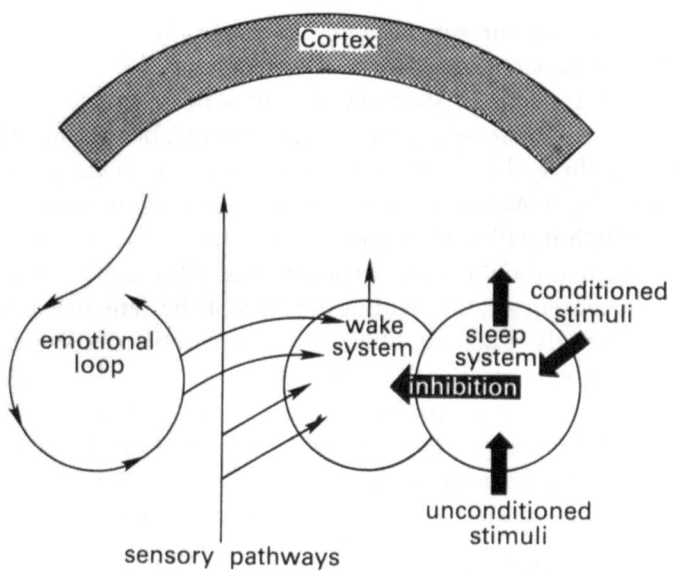

Figure 1. A schematic representation of the sleep–wakefulness balance and the factors which influence it. Wakefulness and sleep depend on the interaction of antagonistic brain stem regions. These are influenced by various factors, and sleep occurs when the factors affecting the sleep region overcome those affecting the wakefulness region (reticular activating system)

The balance between sleep and wakefulness depends on the activity of the brain stem systems – i.e. reticular activating and raphe sleep – which in turn are affected by the balance of the impulses received. These impulses come mainly from peripheral receptors (warmth, light, sound, pain, etc). However, there is a further possible influence from emotional aspects generated in the limbic system. Under normal circumstances this emotional influence is small, but as we shall see later it is an important aspect in many sleep disturbances (e.g. anxiety, depression, conditioned insomnia). Natural sleep occurs when the predominance of factors favouring sleep is added to a quiet reticular activating system (Figure 1). During such natural sleep, a new or different excitation can produce a partial or complete return to wakefulness.

For the state of rapid eye movement sleep to occur, there must be active involvement of the nucleus coeruleus as well as the giganto-cellular tegmental area of the upper portion of the pons. It appears that activity within the giganto-cellular tegmental area is suppressed by the nucleus coeruleus. When the nucleus coeruleus ceases to fire the giganto-cellular tegmental area, freed from inhibition, transmits impulses widely through the brain, and it is thought that these impulses give rise to the phenomena associated with rapid eye movement sleep.

Sleep in Man

Much of our knowledge of sleep in man has been gathered in sleep laboratories, and over the years these studies have contributed considerably to the present day understanding of sleep and of its disorders. Recordings of the electrical activity of the brain, eye movements and muscle activity are often linked to assessments of performance, and the two approaches help us to understand the sequelae of sleep disturbance, and the balance between the efficacy of hypnotics and their possible adverse effects.

A typical array of electrodes used in sleep recording is shown in Figure 2. When only one channel is used a frontal derivation, such as C_4–A_1 is recommended, and either the right or left side may be used as their electrical patterns are generally synchronous. Additional channels from the parieto–temporal (P_1–T_5) and occipital (OzPz–0_3)

15

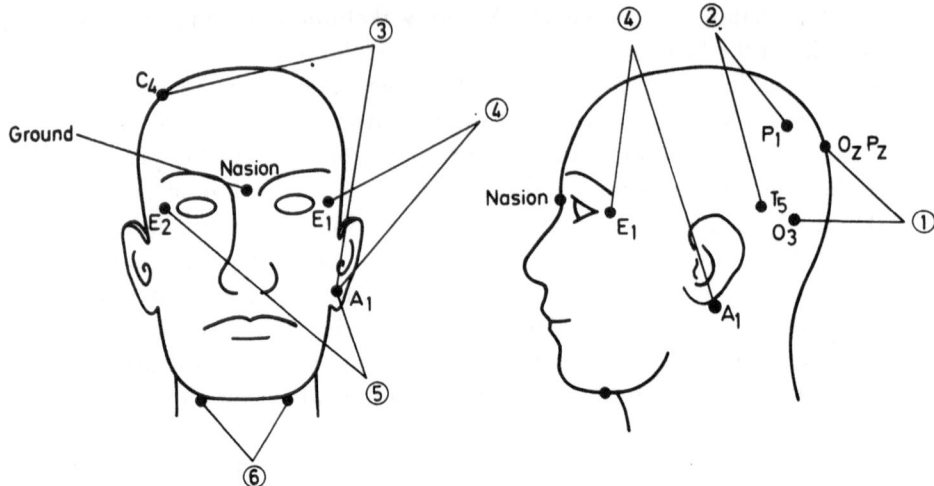

Figure 2. Electrode replacements. The electroencephalogram (e.e.g) is recorded on three channels (C_4–A_1, P_1–T_5 and O_zP_z–O_3), together with two channels of electro-oculography (e.o.g.) (E_1–A_1 and E_2–A_1) and the submental electromyogram (e.m.g.)

are useful, as they help in the definition of particular activity such as the alpha rhythm. This is more obvious from the occipital area than from a single vertex channel. The paper speed for recording is usually 10 mm s⁻¹, as this is the slowest speed which will permit easy identification of the alpha rhythm and of sleep spindles which signify the onset of sleep.

To recognize easily the eye movements which are a feature of rapid eye movement (REM) sleep an electro-oculogram with at least two channels of recording are required. The potentials are recorded between electrodes lateral to the outer canthus of each eye and a reference electrode. This arrangement records eye movements as out of phase deflections. The electromyogram is also used for the recognition of rapid eye movement sleep, as electromyographic activity is at a relatively low level during this phase.

In the clinical assessment of sleep disorders which involve apnoea and paroxysmal movements, recordings of respiratory rate and airflow, together with oxymetry and myography of the anterior tibialis muscles, may also be needed. Hence a complex array of recordings is necessary for the full investigation of sleep and of its disorders.

The electroencephalogram, electro-oculogram and electromyogram are all used to recognize the stages of sleep. The records are read in epochs of 30 second duration, and each epoch is assigned to a single stage. When more than one stage is present in a single epoch, the one which takes up the greater portion of the 30 seconds is selected. Sometimes more than half the tracing may be obscured by muscle activity or artefacts associated with movement of the subject, and then the epoch is scored as movement time. Sometimes the recognition of a stage will depend on the activity of neighbouring epochs.

The waking state is often characterized by the alpha rhythm, but although some subjects have virtually a continuous alpha record, (Figure 3) others may show little or none. The change from waking to drowsy sleep is characterized by slowing of the electrical activity, and by a decrease in the amount, amplitude and frequency of the alpha

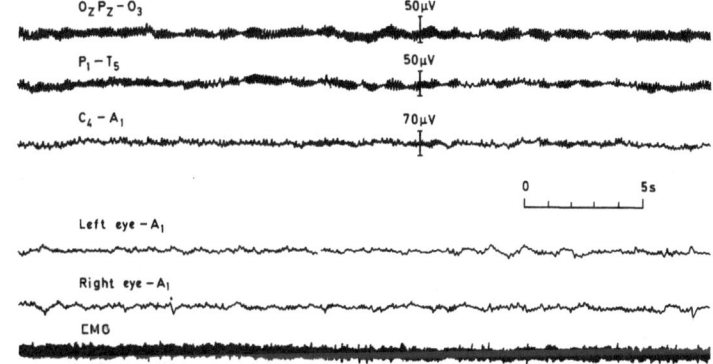

Figure 3. Awake activity in a subject with a dominant alpha rhythm seen most clearly in channels O_zP_z–O_3 and P_1–T_5. E.m.g. activity is sustained and of relatively high amplitude

rhythm. When alpha activity is replaced by relatively low voltage mixed frequency activity, the epoch is scored as stage 1 (drowsy) sleep (Figures 4 and 5).

Drowsy sleep tends to occur in the transition from wakefulness to sleep, or after body movements. It is usually of short duration – a few minutes, and in the later part high amplitude vertex sharp waves may appear. In drowsy sleep, particularly after wakefulness, there are often slow eye movements each of several seconds duration, but rapid eye movements are absent. During wakefulness there is usually a high

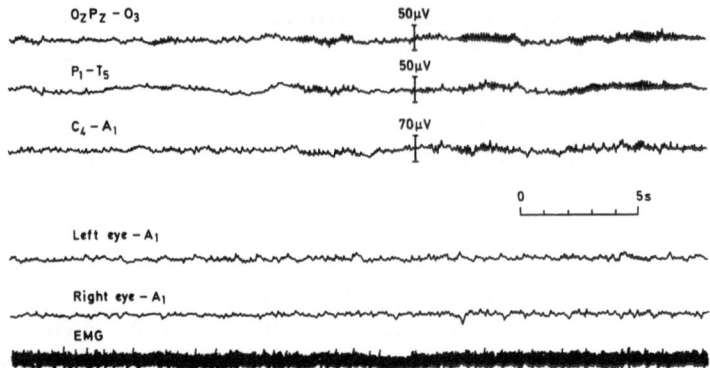

Figure 4. This recording illustrates the transition from awake to drowsy sleep (stage 1). Alpha activity decreases and is replaced by the relatively low voltage, mixed frequency activity typical of drowsy sleep. E.m.g. activity is maintained

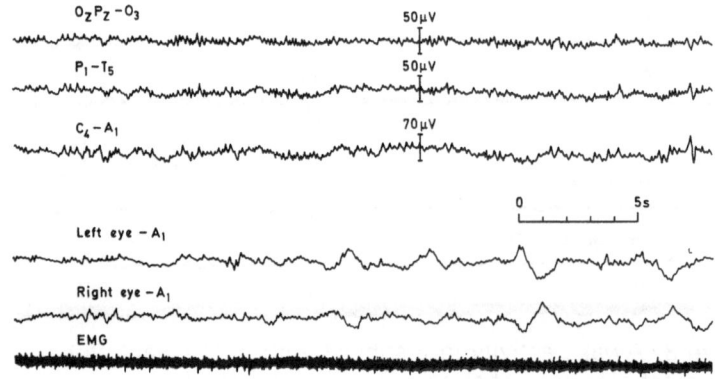

Figure 5. Drowsy sleep (stage 1). The e.e.g. consists of relatively low voltage, mixed frequency activity. Slow eye movements can be seen in the e.o.g.

amplitude electromyogram, but the amplitude of the electromyogram is usually lower during drowsy sleep than during relaxed wakefulness.

Stage 2 sleep is of special importance. It involves spindles and K-complexes, and their appearance is taken as the onset of sleep. A sleep spindle lasts for at least 0.5 second and consists of 6–7 distinct waves of 12–14 Hz. K-complexes are waveforms which have a well defined negative sharp wave followed immediately by a positive component. The total duration exceeds 0.5 s and waves of 12–14 Hz may or may not constitute a part of the complex. K-complexes occur as a response to sudden stimuli, though they appear in the absence of any obvious stimulus (Figure 6).

Stage 2 sleep develops into slow wave sleep when high amplitude waves of 2 Hz or slower with or without sleep spindles appear – stage 3 – (Figure 7), and when more than half of the epoch consists of 2 Hz waves, sleep is scored as stage 4. Although only slightly more than half of the record may contain high amplitude slow waves, most stage 4

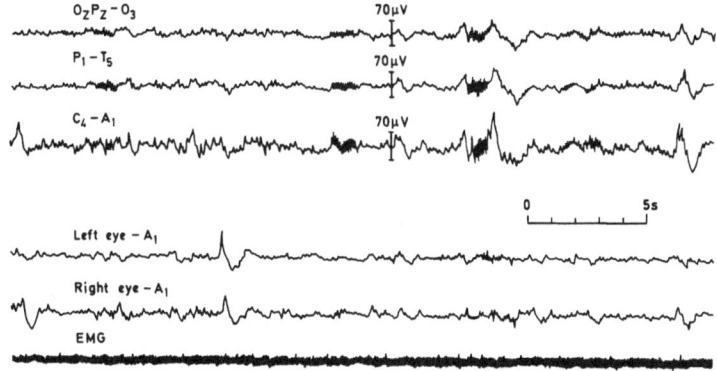

Figure 6. Stage 2 sleep (sleep onset). Stage 2 is characterized by the appearance of spindles and/or K-complexes

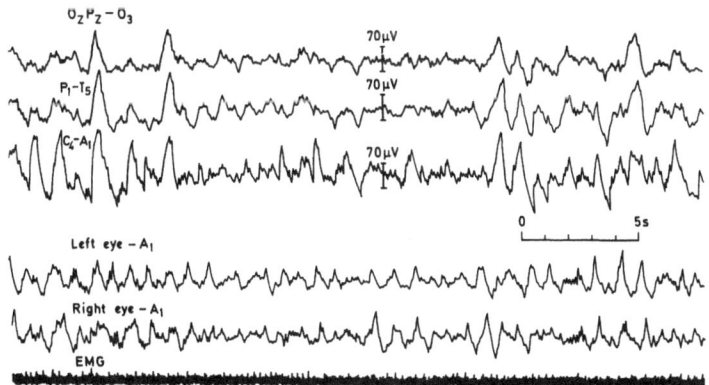

Figure 7. Slow wave sleep – stage 3. This stage is defined by the presence of slow wave activity which occupies at least 20%, but not more than 50%, of the duration of the epoch

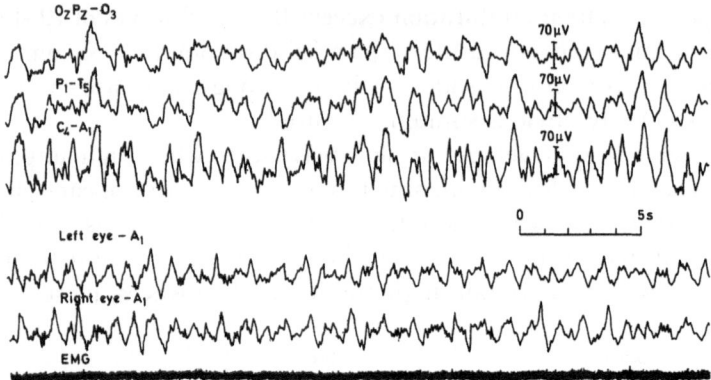

Figure 8. Slow wave sleep – stage 4. This stage is defined by the presence of slow wave activity which occupies more than 50% of the duration of the epoch

sleep has the appearance of being dominated by slow wave activity (Figure 8). The present tendency is to consider stages 3 and 4 as a single category of slow wave or delta sleep, but sometimes it is possible to show discrete effects of drugs on each stage. Stages 1–4 are often referred to as non-rapid eye movement (NREM) sleep.

Relatively low voltage activity together with rapid eye movements indicates rapid eye movement (REM) sleep. The electroencephalogram has some resemblance to drowsy sleep, except that vertex sharp waves are absent. So-called 'saw tooth' waves appear frequently, but

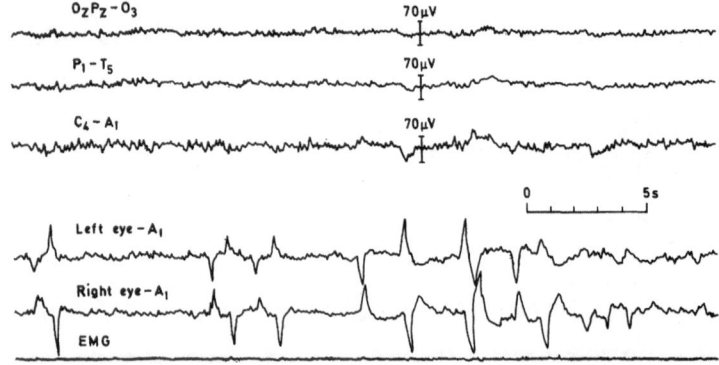

Figure 9. Rapid eye movement (REM) sleep. There is low voltage mixed frequency activity with episodic eye movements. E.m.g. activity is at its lowest level

not always, in vertex and frontal regions in conjunction with bursts of rapid eye movement activity. As with drowsy sleep, there are no sleep spindles or K-complexes. During rapid eye movement sleep the amplitude of the electromyogram is never higher than the level during the preceding sleep stages. Indeed it almost always reaches its lowest levels during this stage (Figure 9).

The Normal Sleep Cycle

In general, the healthy young adult passes from waking through the various stages of non-rapid eye movement sleep before the first period of REM sleep. The normal sequence from waking is through stages 1, 2, 3 and 4, and then returning through 3 and 2 before the first period of REM sleep which occurs about 70–90 minutes after sleep onset. There are then more stages of non-REM sleep before the second episode of REM sleep. The interval from the beginning of one REM period to the beginning of the next has a duration of between 70 and 120 minutes. The mean duration of this cycle changes during the night, and there may be different patterns of change in different age groups.

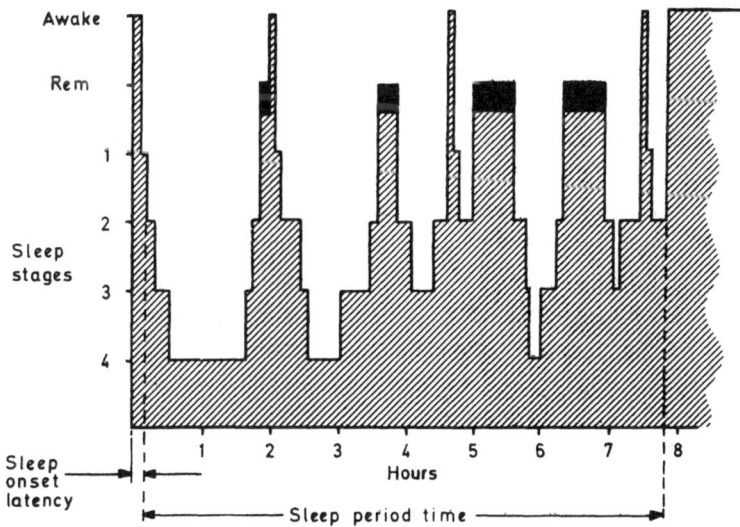

Figure 10. Hypnogram of a young healthy adult which illustrates the sleep stage transitions throughout the night

21

Sleep alters as the night proceeds. In general, slow wave sleep is a feature of the first third of the night and REM sleep is a feature of the last two-thirds. The amount of slow wave sleep before the first period of rapid eye movement sleep decreases with advancing years. The REM episode in the first cycle is shorter than those in subsequent cycles, except in the elderly in whom it may be longer. The pattern of night time sleep in a young adult is shown in the hypnogram (Figure 10), and typical percentages for the whole night are: 50% of stage 2, 25% of REM, 10% of stage 3, 10% of stage 4, and 5% of stage 1.

Circadian Rhythmicity

Sleep and wakefulness vary with time in a periodic and regular manner. This rhythm is normally synchronized with the solar day, though the question arises, as with other rhythms, whether it originates from within the organism or is the result of an exogenous influence. For example, with body temperature the higher value observed in the day and the fall during the night could merely relate to the activity of the individual.

To decide whether a rhythm is endogenous it is necessary to remove time cues. This is difficult with man who is not only influenced by the alternation of light and darkness with its almost constant cycle length, but who has also developed a pattern of rest and activity with regular meals. However, under constant conditions, such as in caves and isolation units, many physiological and psychological rhythms continue to oscillate but with a period length somewhat different from 24 hours. This persistence of rhythmicity shows that circadian rhythms, such as sleep and wakefulness, are endogenous, even though they may be influenced by variations in the environment.

Changes in performance during the day are also an important aspect of the circadian rhythmicity of man. Performance on most tasks rises during the day to a peak or plateau between 1200 and 2100 hours and falls to a minimum between 0300 and 0600 hours when we are normally asleep. This pattern is similar to that of body temperature. However, performance which involves memory falls steadily during the day.

Many factors influence circadian rhythms. If subjects stay awake, the phase of the rhythm tends to drift towards later hours, and as sleep loss continues the level of performance may fall. Further, practice and extra effort reduce, while workload and the stress imposed by the task may increase the amplitude of variation. In the case of performance, the range of oscillation tends to be low for simple tests in highly motivated subjects, and high for complex tasks in those with poor motivation.

It is, therefore, evident that performance over several hours depends on the stage of the circadian cycle with which the period of work coincides. When long periods of work, say 12 hours, begin at noon, performance declines between 10 and 15% of control levels, but when the same work starts at midnight the decrease may be as much as 35%. During the day increased arousal partly compensates for the effects of prolonged work, whereas at night sleepiness may add to the problem. A period of sleep during prolonged work is often of benefit. For example when two 4 hour work periods starting at 2000 and 0400 hours are separated by a sleep, the fall in performance at the circadian trough is reduced.

Conflict of circadian rhythmicity with that of the environment arises when the environmental synchronizers are weakened, disappear completely, or change their period length. This occurs in submarine and space operations, as well as in partial removal from external periodic inputs, as in the Arctic. Rest and activity patterns are also out of phase with the environment in shiftworkers – particularly when night work is undertaken. However, in man the clock hour and daytime-related social activities remain important synchronizers.

Sleep: Age and Gender

Total sleep time is longest in infancy and shortens from around 12 years until the age of 20. It remains constant during adulthood but is reduced during old age, though more time is spent in bed by the elderly when they may be simply resting or unsuccessfully trying to sleep. Awake activity also changes with age. Time spent awake during the night changes very little from childhood to middle age, but then it increases rapidly. Increased wake time in the elderly is not due to an increase in the time to fall asleep, but is due to more frequent and longer awakenings. In this way the sleep efficiency of the elderly may be low. Awakenings may be particularly significant in the elderly as frequent arousals of short duration may be a cause of poor sleep, and may even indicate pathology such as sleep apnoea.

Perhaps the 'most obvious change with increasing age is the decrease in slow wave sleep. This declines throughout life, and the decline is more pronounced in men than in women. Some elderly men may not show any slow wave sleep, though it may still be present in women. However, analysis of the sleep electroencephalogram into sleep stages alone may not reveal the whole story. There may be a reduction in amplitude rather than a decrease in the incidence of activity at this frequency, and the significance of such a change is not fully understood. The proportion of REM sleep remains very constant from adult life into old age at about 20%.

Though women may not be different from men in sleep duration during their twenties, they tend to have a consistently longer sleep time than men over the span of life. The number of awakenings is considerably higher for men than for women, and this difference continues until the seventies. The young adult pattern of sleep is maintained somewhat longer in women.

Variants of the Sleep–Wakefulness Cycle

Short and Long Sleepers

Although there is a general pattern of sleep there are variations of the sleep–wakefulness cycle from individual to individual. A common variation is the length of sleep, and there are short and long sleepers. Individuals whose sleep within 24 hours is substantially shorter or substantially longer than the average amount for their age, and whose sleep is unbroken and normal, without complaints about the quality of sleep, morning alertness, daytime sleepiness or performance, fall into these simple groups. Short sleepers are considered as those who sleep less than three-quarters of the norm, i.e. 6 hours or less. They may even sleep less than 3 hours each day. Long sleepers sleep at least 9 hours a day – possibly as long as 12 to 14 hours, and, as they enjoy their sleep, they may have difficulty in coping with restricted sleep schedules. However, with training many can reduce their sleep by about an hour without any loss of efficiency. In so-called short and long sleepers there is no underlying pathology, though it is said that short sleepers may be 'efficient' and long sleepers may 'worry'.

Advanced and Delayed Sleep Phase Syndromes

Differences of more significance may arise when there is a misalignment between the sleep–wakefulness cycle of the individual and the circadian changes of the environment. Such abnormalities include the so-called delayed and advanced sleep phase syndromes in which sleep onset and wake time are later or earlier than desirable. Sleep tends to occur at the same clock time each day, and there is no difficulty in maintaining sleep once it has begun. The delayed sleep phase syndrome is usually seen in young people. They complain of not falling asleep until the small hours of the morning. They may have problems with getting up, and if their sleep is curtailed by social obligations there may be daytime sleepiness. The advanced sleep phase syndrome is much less common, and does not interfere with daytime alertness. The complaint is that of an inability to stay awake during the evening, and so to maintain sleep until the conventional morning hour.

Less Common Disorders

There are other far less common disorders linked to circadian rhythmicity. There may be a delay each day in the onset of sleep and waking so that sleep and wakefulness occur at a later clock time on successive days. The condition reflects the inherent circadian sleep–wakefulness period around 25 hours which is usually entrained by the 24 hour environmental rhythm. On the other hand certain individuals may have an irregular sleep–wakefulness pattern where there is a loss of anything which resembles a 24 hour rhythm. In this situation the sleep pattern is totally disorganized. Sleep at night is not adequate, and there are frequent daytime naps at irregular times.

Neuroendocrinology

The secretion of many hormones is related to sleep, and in some cases to specific electroencephalographic activity. In most subjects peak

25

plasma levels of growth hormone are seen during the first 90 minutes of sleep. When sleep onset is delayed, the plasma peak of growth hormone is also delayed, and when subjects are awakened for 2–3 hours and sleep is resumed, there is another peak, and smaller peaks which occur during the night tend to be related to slow wave sleep.

The 24 hour pattern of growth hormone secretion varies with age. In the first few weeks of life, growth hormone levels do not vary between sleep and wakefulness, but after the third week of life waking values decrease considerably. Children secrete most growth hormone during sleep and little when awake, and during adolescence both sleep related and day time secretion are greatly increased. There may well be some truth in the saying that children need sleep to grow. In young adults sleep related secretion is somewhat less than in adolescence, and in the elderly it is greatly decreased. This change may be related to the reduced amounts of slow wave sleep. After inversion of the sleep–waking cycle the pattern of secretion of growth hormone is also reversed. The balance of current evidence suggests that it may not be specifically associated with sleep onset, but rather to slow wave sleep.

On the other hand adrenocorticotrophic hormone secretion is lowest in the few hours before and after sleep onset. It increases after 3–5 hours of sleep and reaches its maximum just before awakening. Individual episodes of secretion tend to occur about 10 minutes before an episode of cortisol secretion, and cortisol levels are lowest during the early hours of sleep and highest in the early morning hours. It is difficult to be certain that any relationship exists between cortisol secretion and specific sleep stages. Maximum plasma cortisol levels are normally reached after several hours of sleep, and so a relationship with REM sleep which is highest at that time has been suggested, as most peaks occur in or around REM episodes. Rapid eye movement sleep and cortisol secretion can, however, be dissociated in sleep inversion.

With prolactin there is an initial peak in secretion shortly after sleep onset, with subsequent peaks between 0070 and 0080 hours. During the hour after wakening the levels begin to fall, reaching a minimum around noon. When the hours of sleep are modified, there is an immediate shift. In this sense the secretion of prolactin resembles that of growth hormone, but unlike growth hormone, its secretion does not seem to be related to a specific sleep stage. There is also a general relation of thyroid stimulating hormone secretion to sleep although the initial rise may occur in the hours immediately before nocturnal sleep. If sleep is delayed, thyroid-stimulating hormone levels climb for longer and this may imply that sleep initiates the fall in plasma level.

Chapter 2

Transient Insomnia

Patients with insomnia may complain of inadequate sleep or excessive daytime sleepiness. Some have a chronic problem, but the majority have a temporary difficulty which may be converted into a chronic one by inappropriate management. Patients with chronic insomnia usually have an unremitting history over months or even years. Such patients need careful assessment, as they may have some medical or psychiatric problem or suffer from a specific disorder such as sleep apnoea. Transient insomnia, on the other hand, usually presents with a history of a few days or weeks though the problem may have recurred if linked to particular circumstances. It may arise from a change in the environment, difficulties associated with changes in work or rest or frequently from an acute emotional crisis.

Environmental Factors

Change in the environment can lead to problems with sleep. These include noise, light and temperature – broadly there is no single aspect related to the environment that can be guaranteed to be conducive to sleep or, on the other hand, to disrupt it. There are individual variations of preference for temperature, light and background noise. Variation from the preferred state is a situation likely to produce insomnia, particularly if the change occurs rapidly. The nature of the bed must be taken into account, both in terms of its resilience and covering. A sudden spell of hot weather and the initial experience of central heating can provoke a spell of sleeplessness. So too can a move from the constant noise of a busy metropolitan area to the peace of a cottage in the country.

From this it follows that any situation must be related to that preferred by the individual. As with so many aspects of sleep it is not possible to define arbitrary norms, and practitioners must be careful that they do not concentrate on their own preferences. Another factor which can give rise to insomnia is the ingestion of pharmacologically active substances as food or medicine. When such ingestion (late night consumption of coffee) gives rise to transient insomnia the relationship is usually obvious to the sufferer, but chronic misuse is often forgotten and this question is considered later (p. 41).

Shiftwork

Shiftwork is required for a variety of reasons, and economic considerations have played an important part in the increase of this activity. Work by day and rest at night means that the circadian variations in physiological and psychological functions are in harmony with the routine, but in shiftworkers rest and activity patterns are out of phase with the environment. A single night shift does not change the circadian rhythm of body temperature. Consecutive night work for at least 7 days is required to shift the time of minimum temperature to a point within the new sleeping period, though other physiological functions may not re-entrain at the same rate.

The extent and significance of desynchrony in shiftworkers depends on the individual as well as the work–rest pattern. Shift patterns show considerable variation. Some people choose nightwork permanently, while others prefer rotation between day and night. Some shift systems include weekends and in these abnormal rest and activity patterns may be a permanent feature.

Disturbed sleep is one of the consequences of shiftwork. The night worker rests during the day when the environment does not favour sleep. There is the obvious increase in light and noise levels, and, in turn, these changes may be more easily appreciated during the light sleep of the day. There are also higher ambient temperatures and the surrounding social life may disturb even the most tired morning sleeper. Workers who sleep nearly 8 hours at night may only manage about $5\frac{1}{2}$ hours during the day.

Sleep length associated with early morning shiftwork (0600–1400 h) is close to that obtained with a night shift, and this would suggest that it may not be the night shift alone which causes difficulties. Subjectively, shiftworkers rate the quality of sleep at various times differently. Sleep after the night shift is rated worse than that after other shifts. Sleep before an early morning shift will often be curtailed because of the implications of retiring early, whereas the end of a sleep period after a night shift is more determined by the individual.

It is not only the length of sleep that is important, but also its 'quality'. In shiftworkers there are no differences in sleep onset, but drowsy sleep tends to be increased and stage 2 sleep reduced. In daytime sleep there is a tendency to wake up just before the peak in body temperature, and studies in isolated environments have suggested that when body temperature is low the chance of sleep is higher. The duration of sleep may, therefore, depend in part on its position within the temperature cycle.

Older people may have greater difficulty in adapting to shiftwork. Indeed sleep length and sleep quality in those on night shifts are related to the age of the worker – increasing age being associated with more problems. This often arises around 45 years – even in experienced workers. On the other hand those over 45 years may prefer early morning work, and so a general difficulty in shift working is not necessarily related to age. It is possible that internal de-synchronization of the circadian system is more likely with increasing age, and makes the individual more susceptible to disturbances of the sleep–wakefulness cycle, while 'morningness' may increase with age with earlier peaks in temperature and activity.

Rhythms of body temperature and simple performance follow a similar pattern over a 24 hour period, with a minimum in the early morning hours. However, short-term memory or logical reasoning may have a pattern completely out of phase with that of temperature. Therefore when night work involves 'simple tasks', performance during a night shift may be worse than during a day shift, whereas, this may not be the case for more complex tasks. During a two week period of consecutive night work efficiency gradually changes from a pronounced within shift decrement to a relatively constant level of performance. In general, memory loaded performance rhythms seem to adjust more quickly than others to changes in the sleep–wakefulness cycle.

31

It is not possible to predict whether or not any particular individual is likely to tolerate shiftwork for many years. However, it has been suggested that tolerance to shiftwork is associated with large amplitude circadian rhythms and a slow adjustment to altered schedules. This does not imply that the natural circadian amplitudes differ between tolerant and non-tolerant shiftworkers, but it appears that it is the response of the circadian system to altered work patterns which differs. If this is true, schedules that do not allow the temperature rhythm to adjust to a new 'synchronization' would appear to be preferable, and this would imply that a rapid rotation (change every 2–4 days) is a better choice than weekly rotation.

Although the physiological disturbances associated with shiftwork have been studied in detail, social difficulties are probably the more important for the shiftworkers themselves. It may be possible to plan shift systems which minimize any disturbance, but these may not be acceptable as they may involve undue disruption of family and social life. Indeed in the design of schedules, social needs must be borne in mind.

No single system is optimal for all shiftworkers and for all working conditions, but certain criteria should be considered. Single shifts are preferred to consecutive night shifts, as circadian rhythms are not altered by working a single night. Working for more than 7 nights leads to re-entrainment, but for social reasons most workers prefer a change of shift or rest days after no more than a week. Re-entrainment is therefore not normally possible. Sleep disturbances and sleep deficit over several days should be avoided. To prevent sleep deprivation, an adequate recovery period is desirable. At least 24 hours of free time is desirable after each night shift. Similar problems may arise with early morning shifts. In this case a 24 hour rest is also beneficial, though reorganization of the shift system with later starting times would be an alternative solution.

The length of the shift should be related to the type of work. With light work the shift duration may be extended to 12 h, but it should not exceed 8 h, or even 6 h, when heavy physical expenditure or a high mental workload is involved. The cycle of the shift system should not be too long (i.e. 4 weeks is better than 40 weeks), and a regular system of rotation is preferable to an irregular one. Short cycles and regular systems make it easier for the worker and his family to arrange their social life. In the case of continued shiftwork, it is important to arrange as many free weekends as possible so that a reasonably normal social life is possible.

Because shiftwork cannot be sustained by about 20% of the working population, selection is important. Shiftwork may be contraindicated. New employees living alone, those under 25, and those over 50 years of age should be considered carefully, though experienced and well adapted individuals can in many cases remain in shiftwork beyond the age of 50. A history of digestive tract disorders may lead to problems exacerbated by unusual times of meals or to increased caffeine ingestion and smoking which are common in night-workers. Diabetics and patients with thyrotoxicosis may also find it difficult to ensure regular food and correct timing of medication, and the incidence of fits in epileptics may be increased by sleep reduction. Such patients should be advised to avoid shiftwork if possible.

Transmeridian Flight

Transmeridian flights produce rapid and often large time zone changes. The sudden shift disengages the environmental and biological rhythms, and is likely to lead to disturbed sleep. Subjectively most people complain of tiredness, impaired appetite and a general loss of well-being, while sleepiness is experienced at inconvenient times. There is likely to be difficulty in falling asleep, spontaneous awakening during the night and early awakening in the morning, and when it is the local time to sleep the individual may begin to feel awake. Desynchronization also influences performance. Performance is impaired in the late afternoon and early night of at least the first day after flights in a westerly direction, and in the morning and afternoon after travelling eastwards. The whole syndrome is commonly known as 'jet-lag'.

After westbound flights the amplitude of circadian rhythms may be altered, and sometimes a rhythm may disappear. Re-entrainment of rhythms to the new environment occurs gradually, and for body temperature begins with the first sleep in the new zone. After westward flight the maximum may shift earlier or further than the minimum, whereas the opposite may be true after eastward flights. Although phase shifting is a consequence of the desynchrony between environmental and biological rhythms, some changes may be more related to the general stress of the flight.

In man social cues and timing of meals are important to the speed of re-entrainment. Indeed, deliberately adopting local times for meals accelerates the phase shift. Nevertheless, individuals differ in the ease and the speed of their adaptation, and hence the extent of their jet-lag. Circadian rhythms may synchronize at different speeds, and this is true for psychological as well as physiological functions, although it is not clear how the dissociation arises. As far as performance is concerned it would appear that task complexity is important, and skill with a high memory load may adapt faster than a more simple skill. Using performance as an indicator, resynchronization times for the westward direction vary from between 1.7–6.0 days and for the east-bound direction from between 2.9–11.3 days. In some subjects eastbound travel may require even longer times for complete readjustment.

The most useful approach to the problem is to travel at the right time of day if the airline schedule permits. There is a simple piece of advice for the intercontinental traveller: travel west – travel later; travel east – travel earlier. In both cases, one should try to arrive in time for bed. However, even if sleep loss is avoided during the journey there is the adaptation to the new environment. Adjustment needs time, and some difficulty with sleep for a couple of days is likely. A period of rest is needed before working and making decisions, and the duration of rest should be related to several factors, including the duration of the journey, the number of time zones crossed and the times of departure and arrival.

Some journeys from north to south without a significant time zone change require no particular period of rest, though such travel should be followed by rest if the departure was late and sleep was disturbed. Recovery after transatlantic travel depends on the direction and on the time of departure. As an example, leaving Montreal at 2100 hours would require a day's rest in London, but leaving London for Montreal at 1500 hours would not require any rest the next day. On the other hand departing from Montreal at 2300 hours for Karachi should be followed by 2½ days rest, and leaving Montreal for Sydney at 1000 hours needs a similar period of time.

Emotional Crisis

Difficulty in getting to sleep, frequent awakenings and early awakening are commonly due to psychological factors (Figure 11). Indeed it has been suggested that psychological factors play a part in as high a proportion as 80% of the patients who visit their practitioner with transient insomnia.

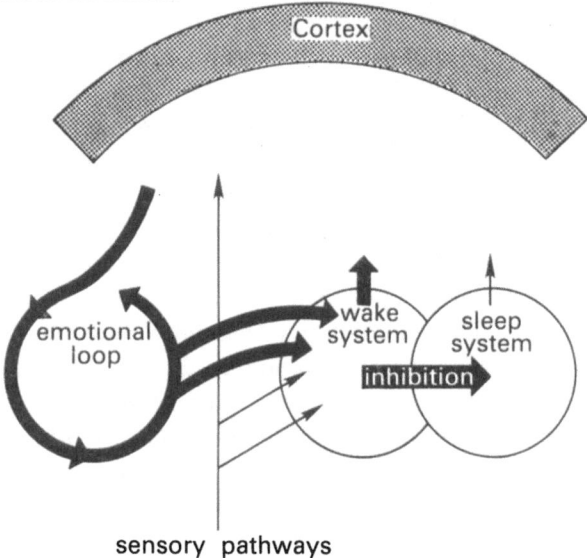

sensory pathways

Figure 11. Compare with Figure 1. A schematic representation of heightened emotions leading to wakefulness (insomnia)

Any individual who is under pressure, due to a difficult domestic or work problem, often finds that difficulty in getting to sleep is among the first symptoms of the inability to cope, and transient insomnia is an indication that stress has become distress. Moreover, the mere fact that the person finds difficulty in achieving the desired quantity and quality of sleep means that he has more time to be awake thinking about the problem and this sets up a vicious cycle. Even when blessed sleep does come, its quality in the anxious person does not appear to be appropriate. People who sleep under stress wake up exhausted, knowing that in some way their minds have not rested.

While acute anxiety states can occur as a primary phenomenon they are rare and most anxiety states are the result of external stress. In many instances the cause of the morbid anxiety will be apparent and simple. It results from the stress of the emotional conflict within the individual between his needs, hopes and aspirations and the realities of the circumstances thrust upon him by life. The circumstances of his employment, finance and domestic life are fertile fields for such conflict, and frustration and shattered pride, thwarted ambition and disappointment may generate powerful stresses. An excess of responsibility with insufficient support from friends and associates can produce a major degree of morbid anxiety in those whose psychological make-up is vulnerable to stress of this kind.

Viewed in terms of quality as well as quantity the type of stress which causes an anxiety response will be highly selective, and will vary from one individual to another. One individual will be subjected to immense problems at work and shrug them off without difficulty only to succumb to a minor marital discord, while his neighbour who has overcome the loneliness of bereavement, suffers from a panic state at the thought of acting as chairman of a business meeting.

There is difficulty in predicting which individuals exposed to stress of this level will show psychosomatic disorders and which will suffer from insomnia. Certainly there is no relationship to the type of stress involved. Moreover, imagined problems are equally likely to lead to insomnia as are real problems. Studies in practice have established that emotional distress, sufficient to produce clear clinical manifestations is found in up to 30% of consultations. Since insomnia is such a common feature of such disorders, it is scarcely surprising that transient insomnia is such a common symptom in the practitioner's consulting rooms.

Chapter 3

Chronic Insomnia, Sleep Pathology and the Parasomnias

Chronic Insomnia

It is difficult to identify the stage at which transient insomnia can be said to end and chronic insomnia to begin. The majority of those working in the sleep disturbance field would insist on at least some weeks (say 3–4) of continuously disturbed sleep, but some would suggest a much longer period. The question is, however, largely academic because in practice the patient usually presents with the complaint of poor sleep either within a matter of days or after some months.

Among those with prolonged difficulties in sleeping we can distinguish three classes, primary chronic insomnia, secondary chronic insomnia and those with sleep difficulties which are part of a sleep disorder.

Primary Chronic Insomnia

Although most chronic insomnias are associated with medical, psychiatric or behavioural problems, some patients have persistently poor sleep in the absence of obvious pathology. It may vary from mere restlessness to continued wakefulness. Chronic insomnia without obvious reasons is little understood either in terms of the nature of the illness or in its management. The condition involves disturbed sleep during the night and loss of well-being during the day. Chronic insomnia may involve two distinct symptom complexes – one with pronounced daytime sleepiness, and the other free of daytime sleepiness but associated with impaired mood and well-being. The aetiology of primary chronic insomnia is uncertain, though two possibilities are worthy of consideration. The patient may have need of more sleep or of a different pattern of sleep or alternatively, the sleep of the patient may involve abnormalities which are not yet appreciated.

As far as the first possibility is concerned it must be stressed that a relatively small reduction in the duration of sleep can lead to a substantial increase in daytime sleepiness – at least as judged by the multiple sleep latency test which assesses the speed at which individuals fall asleep during the day. As far as the second possibility is concerned it is well established that abnormalities in sleep as confirmed by electroencephalography – such as persistent delay to the onset and frequent nocturnal awakenings – occur only in a small proportion (about 10%) of patients who have chronic insomnia without any obvious cause.

Indeed recent research has raised the question that more subtle changes in sleep may be involved in the phenomenon of chronic insomnia. There may be frequent arousals of only a few seconds – 'mini arousals'. They may not be appreciated by the patient and would not lead to scoring an 'awakening' in all-night sleep electroencephalograms. Such changes in sleep which are not identified by current recording methods may lead the physician to the conclusion that the patient slept soundly throughout the night. Other possible changes in sleep which may be associated with the complaint of chronic insomnia are fewer sleep spindles and awakening at regular intervals just after the onset of each period of rapid eye movement sleep.

Chronic insomnia without obvious cause and without obvious signs can easily lead to the label of malingerer or 'neurotic', but there should be caution in such a diagnosis. Further, the elderly should receive very careful attention as minor alterations in sleep – particularly mini arousals, may be of clinical significance.

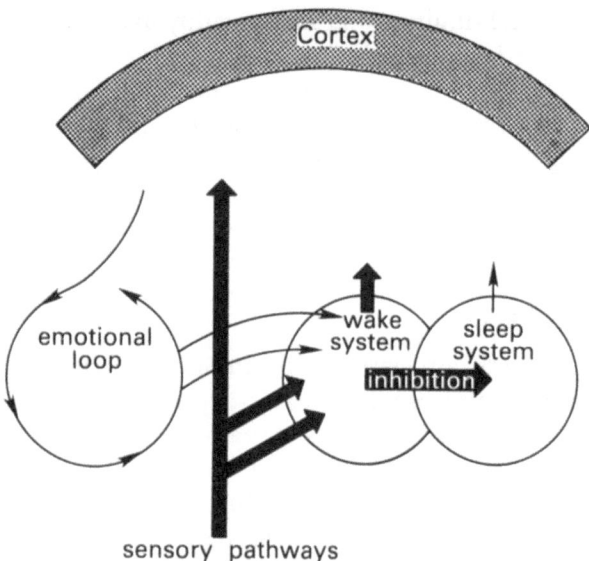

Figure 12. Compare with Figure 1. A schematic representation of painful stimuli producing wakefulness (insomnia)

Secondary Chronic Insomnia

One of the commonest causes of chronic insomnia is pain (Figure 12), for example from arthritis, but almost any medical condition may be associated with insomnia. Sleep may be impaired with injury, infections and degenerations of the central nervous system, as well as with some forms of epilepsy. Nocturnal migraine sufferers may wake from headaches which appear to develop during REM sleep. Hyperthyroidism and uraemia are associated with short fragmented sleeps, and patients with hyperthyroidism may also show excessive slow wave sleep. Hypothyroidism may be associated – though rarely – with excessive daytime sleepiness and the electroencephalogram may show less slow wave sleep. Ischaemic heart disease frequently results in sleep disturbance. Patients who awaken in the night with angina usually wake during a REM period. Cardiac arrhythmias also occur more often in REM than during NREM sleep.

Chronic insomnia may also develop from inadequately treated transient insomnia, due to the patient's failure to seek medical advice, inaccurate diagnosis or inappropriate therapy. In these patients the causes of chronic insomnia are essentially those of transient insomnia except that the self-limiting and more readily diagnosed and treated causes will have been excluded.

There is, however, a clear indication that a substantial proportion of the patients in the general population who experience prolonged difficulty in achieving a satisfactory level of sleep have a psychiatric cause for the difficulty, and that this is commonly either a recurrent anxiety or an undiagnosed depression.

Psychiatric Causes

Affective disorders associated with insomnia include the major depressive and bipolar disorders (manic-depressive) in which there is a current and chronic inability to maintain sleep with early morning awakening. Such an origin for sleep difficulty may be easily missed in the elderly. With both monopolar and bipolar endogenous depression there is a shortened latency to REM sleep, particularly in the elderly, reduced slow wave sleep and repeated awakenings, although in a few patients with bipolar depression sleep may be longer with excessive daytime naps. On the other hand, patients with exogenous depression who complain of insomnia do not usually have a shortened latency to REM sleep, though their sleep is fragmented with reduced REM and slow wave sleep.

Depressive illnesses are normally self-limiting and of relatively short duration, and so the sleep disturbances may be transient. However, some depressions run a more prolonged course, and in such patients chronic sleep difficulties are encountered. In some patients the depressive aetiology may not be recognized and inappropriate long term therapy with benzodiazepines may be given with poor results. Recognition and treatment of the true cause leads to rapid improvement.

Another cause of chronic insomnia, and the one which is believed by many to be the most common cause of all, is chronic or situational anxiety. The nature and cause of the anxiety state which leads to insomnia has already been considered. In theory the reduction of the distress by the use of anxiolytics with temporary relief of the insomnia enables the patient to appreciate and solve the problems that caused the original distress.

In practice however, the causes of many cases of the situational anxiety are incapable of ready solution. The distress stems from a difficult environment – at work, at home or in the family – or from a combination of two or all of these. Sociologists envisage solutions for such problems, but practising physicians appreciate that this is often virtually impossible.

It is for this reason that many of the cases of transient insomnia that arise from situational anxiety follow an intermittent course. The insomnia recurs at regular intervals as the level of distress crescendos. This often presents as a chronic insomnia, and these patients are among the most difficult to manage. For a proportion, the only hope of maintaining a life in society is by the intermittent but prolonged use of sedatives. By the very nature of their complaint many of these patients may become dependent, and drug therapy must therefore be considered carefully before it is started. The difficult question of the management of these patients is considered later.

Patients with psychotic illnesses may also have severe insomnia, and the possibility of schizophrenia or pre-senile dementia must always be borne in mind. Patients with anorexia nervosa tend to have disturbed sleep, particularly in the middle third of the night and wake early. This is unrelated to mood but probably to the severity of the nutritional disturbance.

Pharmacological Causes

Among other secondary causes of insomnia must be considered the effects of pharmacologically active chemicals. Not all such substances are prescribed by the practitioner. Indeed, large amounts of such substances are obtained from dietary sources. Thus the sustained use of stimulants including quite normal amounts of caffeine often taken late in the evening, may give rise to disturbed sleep (Figure 13). Poor nocturnal sleep may in its turn increase the tendency to take stimulants during the day to maintain alertness. This perpetuates the problem in individuals dependent or habituated to stimulant drugs, such as appetite suppressants. Withdrawal may present as daytime sleepiness, frequent napping and long periods of nocturnal sleep.

The ingestion of even a small amount of alcohol also tends to alter sleep. Sleep onset may be shorter, but there may be more awakenings and stage changes (Figure 14). It also tends to depress REM sleep. The chronic alcoholic awakens many times during the night, and has

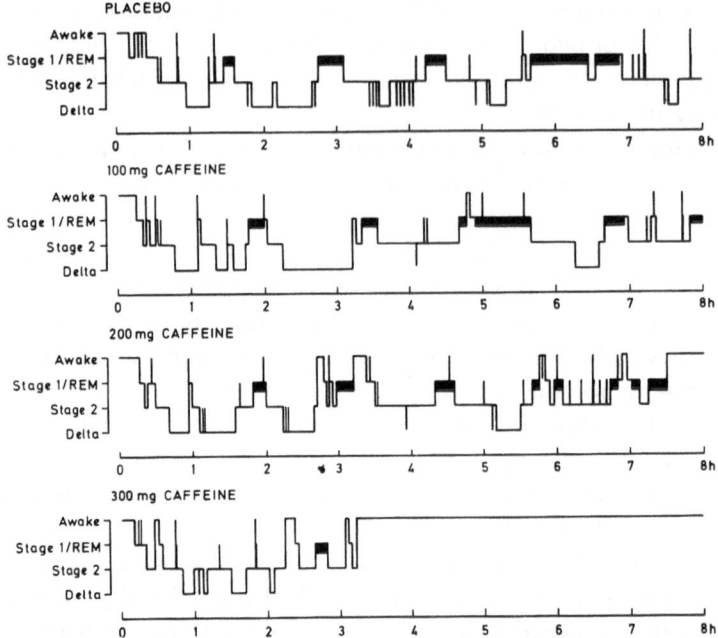

Figure 13. Effect of caffeine on the sleep of a young healthy adult. Sleep structure is not changed, but with increasing doses there are more awakenings. With 300 mg there is sustained wakefulness after a short period of sleep

more stage changes with little or no slow wave sleep. Time in bed may be increased and the patient may experience excessive daytime sleepiness with numerous naps. Insomnia is also a feature of withdrawal from alcohol, and some patients may show continuous REM activity. It must also be remembered that even after a long period of abstinence unusual sleep patterns may persist in the alcoholic who may still complain of difficulty in falling asleep. In such patients the use of any hypnotic is contraindicated for the risk of dependence is high.

Sleep disturbance is associated with tolerance to or withdrawal from other central nervous system depressants, e.g. hypnotics. With continued use, drugs may become less effective, and lead to higher doses. With the chronic use of an hypnotic sleep may be disrupted with frequent awakenings lasting for 5 minutes or more (Figure 15). This

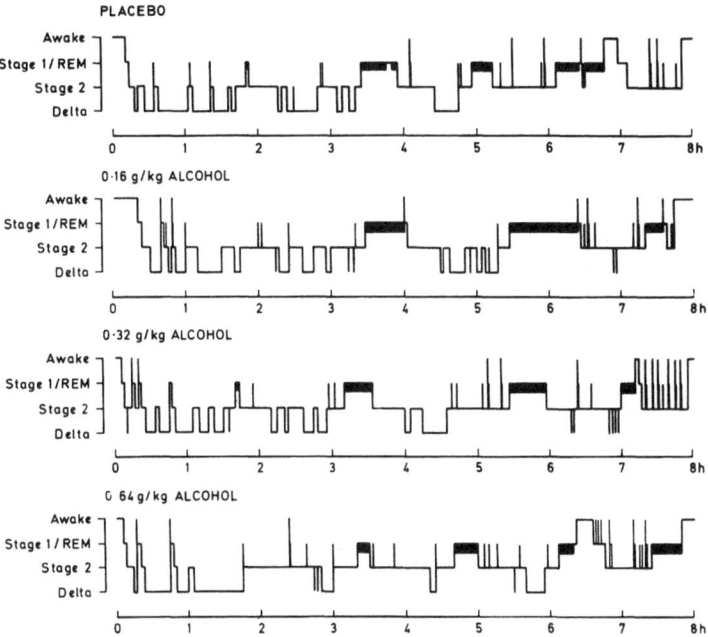

Figure 14. Effect of alcohol on sleep. Small doses of alcohol may disturb sleep, with delay to rapid eye movement sleep and frequent arousals during the latter part of the night

may be a particular problem during the latter half of the night. When the drug is suddenly discontinued severe sleeplessness may supervene, perhaps accompanied by the general features of a withdrawal syndrome. Further, narcotic abuse must not be overlooked.

There is also the possibility that long term therapy with benzodiazepines may lead to unwanted effects when the treatment is discontinued. With benzodiazepines which have long-acting metabolites, the reappearance, after an abrupt discontinuation, of pre-treatment symptoms of anxiety and insomnia may be slower than that with short-acting compounds, though this may not always be the case. There is at present some uncertainty concerning the nature of such effects – referred to as rebound insomnia and rebound anxiety – though they would appear to be related to the repeated use of high doses for unnecessarily long periods of time.

43

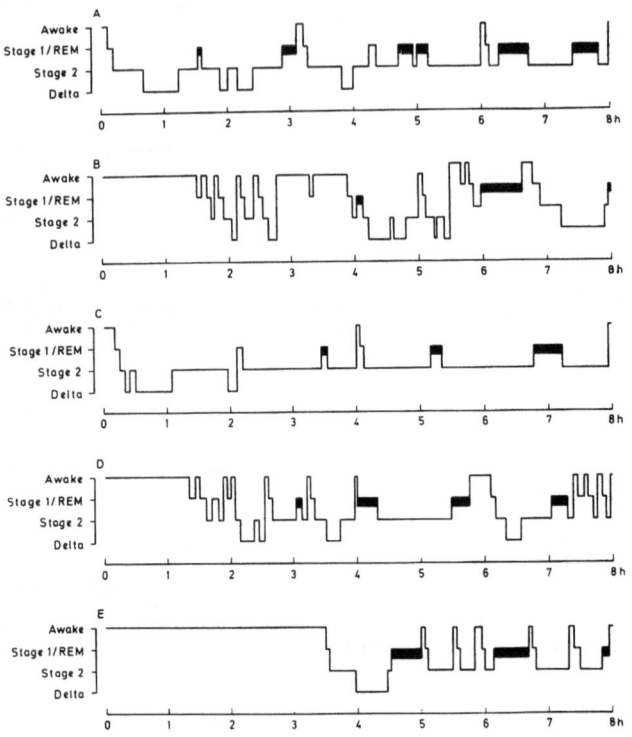

Figure 15. Hypnograms in a young adult with insomnia on 100 mg secobarbital. A) Hypnogram of a young healthy adult. B) Sleep of the young insomniac without medication. There is a delay to sleep onset and there are frequent and prolonged awakenings. Rapid eye movement sleep is also delayed and its duration is reduced. C) During the first night on medication (secobarbital 100 mg) the pattern of sleep onset is normal, but there is a continued delay to rapid eye movement sleep. However, the frequent and prolonged awakenings seen before medication have disappeared. D) After five nights on the drug the pattern of sleep seen before medication has returned. There is a long latency to stage 2 sleep and there are frequent awakenings. E) Sleep during the first night after withdrawal of the medication. There is a very long latency to stage 2 sleep with frequent awakenings.

Reproduced from Hauri, P. (1977), *The Sleep Disorders*, by kind permission of the Upjohn Company, Kalamazoo, Michigan

Sleep Pathology

Narcolepsy

With narcolepsy the patient suffers from excessive daytime sleepiness, and one or more of three other well established symptoms. Disturbed nocturnal sleep may also be a problem. About one in ten patients with narcolepsy suffer from the complete tetrad of symptoms, and those related to wakefulness are the most frequent. The primary and most disabling feature is drowsiness with short periods of daytime sleep, which may sometimes be prevented by concentrating on staying awake. They occur at inappropriate times and last for 10–15 minutes, though if the patient is resting they may sleep for a couple of hours, and they may or may not awaken refreshed. The attacks may occur with or without warning, and are common in situations which provoke drowsiness.

The most common auxiliary symptom is cataplexy, which appears after sleep attacks are well established. It occurs in some form or another in at least two-thirds of patients with narcolepsy, and may occur many times a day, or once a week or even less. There is a sudden decrease or abrupt loss of muscle tone while the patient remains fully conscious. This may be generalized or limited to certain muscle groups, and may result in transient weakness of the jaw or in extreme cases complete loss of muscle tone with postural collapse. The deep tendon reflexes arc lost and the II-reflex is absent. An attack may last for only a few seconds, and is frequently triggered by exercise or by expressions of emotion such as laughing or crying. It can pass into rapid eye movement sleep, and it is thought that the underlying mechanism in narcolepsy may involve a disturbance of the systems which control wakefulness and rapid eye movement activity.

Sleep paralysis, hypnagogic and hypnapompic hallucinations are other manifestations, and occur while the subject is falling asleep or on waking. Recordings during these show REM sleep. In sleep paralysis the patient feels he cannot move any muscles except those controlling the eyes, and this is often accompanied by intense fear and by hypnagogic hallucinations. Respiration is not affected, and the paralysis can be terminated by vigorously moving the eyes or by being touched. It lasts from a few seconds to several minutes. Hypnagogic hallucinations are vivid, frightening auditory or visual hallucinations

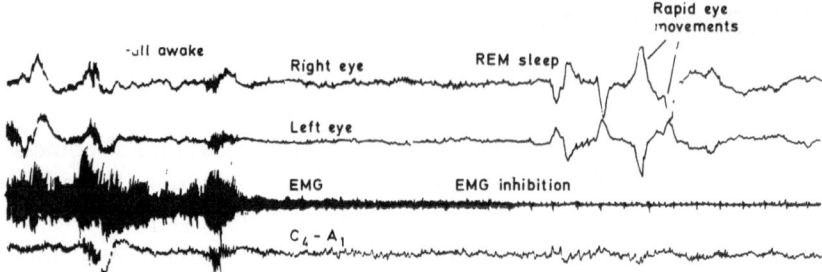

Figure 16. Recording during sleep onset in a narcoleptic subject. The immediate appearance of rapid eye movement sleep (rapid eye movements and loss of e.m.g. activity) is characteristic of this condition. Reproduced from Hauri, P. (1977), *The Sleep Disorders,* by kind permission of the Upjohn Company

experienced when half asleep, and they often occur during sleep paralysis. Sleep paralysis and hypnagogic hallucinations are each present in about a quarter of patients with narcolepsy.

Narcoleptics, particularly those in whom excessive daytime somnolence is accompanied by cataplexy, often have a rapid eye movement period immediately at sleep onset (Figure 16), and sleep may be disturbed with awakenings and body movements, excessive drowsy sleep and little slow wave sleep. Some patients have excessive REM sleep, though usually the duration is within normal limits. A REM period within 10 minutes of sleep is usually considered as evidence for narcolepsy, and rapid eye movement sleep may also be observed during the onset of daytime sleep. The diagnosis is made when sleep onset involves rapid eye movement activity together with sleep attacks and an auxiliary symptom.

The incidence of narcolepsy is probably between 1 and 2 per thousand of the population, and it usually presents itself during the second or early part of the third decade, even though there may be a genetic basis. There is no consistent psychopathology associated with the syndrome. However, the patient may be considered as lacking motivation and interest in his work, and often the history of a narcoleptic will include episodes of a disciplinary nature. Treatment is somewhat uncertain though excessive daytime sleepiness may be alleviated by stimulants, and if the cataplexy is of a serious nature, tricyclic antidepressants without sedative activity have been used.

46

Sleep Apnoea

The hypersomnia – sleep apnoea syndrome is found almost entirely in middle-aged and elderly males. Between the ages of 40 and 60, its incidence may reach 5% of the population. It is characterized by excessive daytime sleepiness (hypersomnia) and frequently recurring apnoea during sleep. The syndrome is usually suspected because of the history of loud intermittent snoring. To diagnose the syndrome at least 10 apnoeic episodes (cessation of airflow at the level of the nostrils and mouth lasting at least 10 seconds) must be observed an hour, in both REM and NREM sleep. Some episodes must occur repetitively in NREM sleep during a 7 hour nocturnal sleep period, for apnoeic episodes may occur at sleep onset or accompany bursts of rapid eye movements in normal individuals.

The condition has been separated into two broad complexes. In central apnoea there is an absence of respiratory effort with cessation of diaphragmatic movement, though the upper airway remains open even though there is no airflow. In the other variety upper airway obstruction is present. The airway is closed and there is excessive respiratory effort. The condition may be associated with the development of hypertension with cor pulmonale, and is often associated with severe cardiac arrhythmias.

Because of disturbed nocturnal sleep and hypoxaemia patients complain mainly about excessive daytime sleepiness, and may take frequent though unrefreshing naps during the day – often at

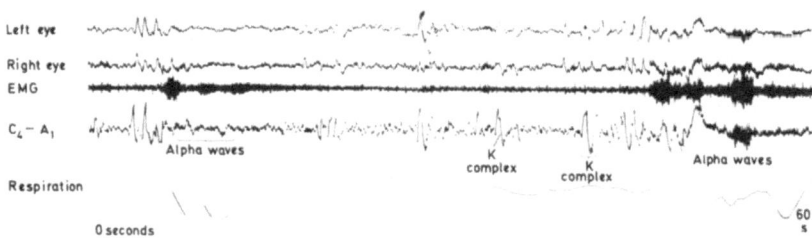

Figure 17. Sleep apnoea. The patient is awake during the period of alpha activity. As he falls asleep (appearance of K-complexes and relaxation of the e.m.g.) breathing is arrested. When breathing restarts there are e.e.g. signs of arousal (alpha waves). Reproduced from Hauri, P. (1977). *The Sleep Disorders*, by kind permission of the Upjohn Company

inappropriate times. The football fan may fall asleep at the match and the teacher may fall asleep in front of the class. Obesity, in particular, and possibly depression may be associated with the condition. Nearly all patients are heavy snorers and the diagnosis must be considered in a patient who snores and complains either of excessive daytime sleepiness or insomnia. Heavy snoring may have been present for many years before the development of the condition. Investigation should include both sleep studies and respiratory function tests during sleep. Electroencephalographically, patients may have short sleep latencies and wake up several times during the night (Figure 17). During the day there is sleepiness and the patient takes long unrefreshing naps. There may be mechanical abnormalities of the soft palate and jaw or even neurological disorders. If these cannot be corrected, or there are no mechanical difficulties, a tracheostomy which is opened during sleep may be necessary in those patients with the obstructive form. Weight reduction and sleeping on the side may help some patients. The possible usefulness of clomipramine is being investigated. Sleep apnoea is worsened by alcohol and by the benzodiazepines.

A related disorder, the alveolar hyperventilation syndrome is a disturbance of sleep due to ventilatory impairment. Ventilatory studies reveal unresponsiveness to chemical control of ventilation during wakefulness and sleep, though pulmonary function tests are normal. During sleep the tidal volume decreases with hypercapnia and hyperoxaemia.

Nocturnal Myoclonus (Periodic Leg Movements and 'Restless Legs Syndrome')

In nocturnal myoclonus, common in the elderly, highly stereotyped leg twitches repeat themselves every 20–40 seconds during sleep. Recordings from the right and left anterior tibialis muscles usually show bursts of activity, and the episodes last from 5 minutes to 2 hours and alternate with normal periods of sleep (Figure 18). In some individuals there are complaints of insomnia though many who show pronounced nocturnal myoclonus do not complain of any sleep disturbance. The condition is quite distinct from the startle movements experienced frequently by many people as they fall asleep.

Many patients with periodic leg movements when asleep also

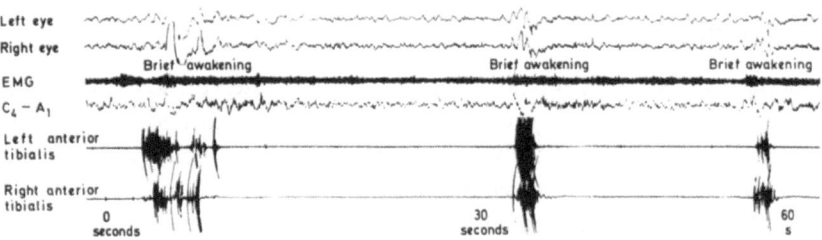

Figure 18. Nocturnal myoclonus. The sleep e.e.g. indicates stage 2 sleep. There are periodic twitches, every 20–40 seconds in the tibialis muscles, and each is accompanied by a brief awakening. Reproduced from Hauri, P. (1977), *The Sleep Disorders*, by kind permission of the Upjohn Company

complain of restless legs before falling asleep. There are uncomfortable and disagreeable sensations of cramp deep inside the calf muscles. The sensations may be ameliorated by movement of the legs but this prevents sleep onset. Some such patients are helped by the muscular relaxation induced by benzodiazepines. Motoneurone disease, iron, calcium and vitamin E deficiencies, as well as chronic uraemia, are reputed to be associated with the condition.

Parasomnias

The parasomnias consist of a group of sleep disorders which involve unusual behavioural and/or physiological events, though the sleep–wakefulness continuum itself is not abnormal. The parasomnias are seen most often in children though they occur in all age groups. They are usually considered as separate disorders though sleep terrors, sleep related enuresis and sleep walking have been considered together as disorders of partial arousal, and since they are related to slow wave sleep appear in the first few hours of sleep. On the other hand bruxism (tooth grinding) does not appear to be related to any specific sleep stage.

Somnambulism

Somnambulism or sleep walking is a sudden interruption usually of slow wave sleep which involves simply sitting up in bed or walking around without being fully aware. However, the behaviour of the individual may be quite complex and involves day-to-day matters such as opening doors and climbing stairs. The whole episode may last up to a quarter of an hour, and during the process the individual resists being aroused though the process may be terminated by a spontaneous arousal. Somnambulism occurs in children and adolescents and may be considered a normal event as it occurs in one-seventh of all children. However, there may be psychological problems and temporal lobe epilepsy must be excluded. When it occurs in adulthood it cannot be considered a normal process as there is also associated psychopathology.

Pavor Nocturnus (Sleep Terror)

An episode of sleep terror involves a sudden arousal during slow wave sleep, i.e. in the first third of the night rather like somnambulism, but it is initiated by a scream and accompanied by agitation and even panic with autonomic changes such as sweating and rapid respiration. It also occurs predominantly in children but may occasionally be found in adults. If the attacks are frequent in children or persist in adults there may be associated psychopathology. It is important to differentiate sleep terrors from nightmares.

Nightmares

Nightmares occur during the later two-thirds of the night as they are related to REM sleep, and the individual – contrary to the person with a sleep terror episode – recalls a vivid experience which may threaten his well-being.

Sleep-Related Enuresis

Sleep-related enuresis is rather like somnambulism and pavor nocturnus as it occurs during slow wave sleep, and so during the first third of the night. It is associated with a brief arousal and some confusion. Enuresis may involve the persistence of bed wetting from infancy into childhood and its reappearance in a child over 3 years needs investigation. It must be differentiated from that which may arise due to organic conditions.

Sleep Related Enuresis

PART II

Management of Sleep

Chapter 4

Sleep Hygiene

For patients who feel that they have unacceptably poor sleep, the experience can be devasting, and it is well described by Oswald.

'Lying awake. Tossing and turning. Mind dwelling on the same eternal problem. First one solution. Then another. Ferment. Back again to the impossibility, the insolubility. Returning, diverging. Fears and possibilities. Endless circling. It happens to everyone at times of life. They are tired, they want to sleep, but oblivion will not come'.

With such a picture, and there are few of us who do not experience it from time to time, it is not surprising that the subjective effects of insomnia can be devastating. Hence insomnia is one of the most common symptoms for which patients seek advice from their general practitioners. However, automatic resort to the use of a sedative is undesirable. Their use may be self-perpetuating and the practitioner should try other methods for re-establishing the normal sleep pattern once he has confirmed that there is no physical cause for the insomnia.

Behaviour which leads to sleep is described by the general term 'sleep hygiene'.

Normal Aspects of Sleep Hygiene

The Expectation of Sleep

As has been explained earlier, the pattern and length of sleep is not uniform. Some people sleep for only 5–6 hours each night and feel refreshed, while others take not less than 9 hours each night and remain tired. Despite this, many patients feel deprived of sleep if they do not achieve their expectations of 7–8 hours sleep per night. As age increases the length of sleep is reduced and is often taken in short naps. An important aspect of sleep hygiene is to have a realistic expectation of sleep, and to accept that constant tiredness during the day is the only clear indication of inadequate sleep.

Patients with reduced sleep needs can be reassured that they are fortunate, for they can enjoy a greater proportion of the 24 hours awake. It is useful to find a method of relaxing during the wake periods of the night, e.g. reading by a well shaded bedside lamp and listening to a small transistor with an earphone.

The Bed and Surroundings

It is often useful to mimic the conditions that existed when sleep was good. Some are most comfortable on a hard bed, others on one which is soft and yielding. Excessively hard surfaces may cause more body movements and a lighter level of sleep with more awakenings than do soft surfaces. However, this may not apply to every individual. In a double bed with a partner a change in the side on which they lie may be all that is necessary. On the other hand, sleeping in separate beds which improves the depth and quality of sleep may be the answer.

Noise may be another factor that leads to a change in sleep. The amount of noise that can be tolerated before sleep becomes disturbed varies markedly from one person to another, but in everyone there is a variation in noise tolerance which depends on the level of sleep. Drowsy sleep is disturbed by the least noise while the stimulus which leads to disturbance is greatest during slow wave sleep. The situation in rapid eye movement sleep is variable. The sound level to produce a disturbance can be high if the sound is such that it is incorporated in the dream content, low if it conflicts with the dream.

Sensitivity to noise while asleep tends to be greater in women than in men, and they waken more easily with unexpected noises – perhaps due to the inherent need for maternal awakening. The influence of noise also depends on the significance of the message that is imparted to the sleeper. Thus a mother can, for example, sleep peacefully through a thunderstorm, but wake with the first whimper from her infant in the next room.

Sensitivity to noise while asleep also increases with age despite reduced auditory acuity – perhaps due to the reduced depth of sleep as age increases. Taken broadly the depth of sleep is less in noisy surroundings and reducing the noise level produces a sounder sleep. On the other hand there is individual variation and unfamiliar noises are the most disturbing. There is a need for trial and error if noise is a problem, but some form of sound muffling, for example by ear plugs, is on the whole helpful.

Since periods of light sleep are most common towards the morning, it follows that the level of tolerated noise tends to be lowest later in the night. The sound of the early morning delivery of milk may be particularly disturbing. Since this also coincides with the period at which the bladder becomes distended, early morning awakening can result from these combined disturbances.

The same variation is seen in the influence of light on sleep. Many people find that light disturbs their sleep, but some adapt and may even need it for good rest. This is particularly true of young children. In some instances it may relate to a fear of the unknown in the dark, while in others it may be a conditioned response to the presence of the light. When a light is preferred it should be kept at low intensity, and should provide a general illumination rather than shine directly on the eyes. A particular problem arises in the elderly. Sleep is often already disturbed, but even so there are advantages in arranging that the bedroom be lit so that falls are less likely. On balance encouragement to adapt to a low intensity of general illumination is desirable.

Environmental temperature is another feature that can produce sleeping difficulties. As with all environmental factors there is no ideal temperature for everyone. The widely held view that sleeping in a cool room improves the sleep does not stand critical analysis. Nevertheless, it has been shown that sleeping in too high a temperature (above about 24°C – 75°F) does reduce the quality of sleep. Both rapid eye movement and slow wave sleep are reduced, there are more periods awake and a greater level of movement during sleep. Gradually reducing the room temperature to 12°C increases the unpleasant and emotional aspects of dreams. A temperature that is equitable to the patient is usually most effective, though a comfortable sleep is achieved more frequently if the room temperature is reduced rather than increased.

A recent problem is the appearance of insomnia when central heating is installed. The person does not appreciate the increase in the general environmental temperature and retains too many bedclothes. The amount of covering is another vexed question to which there is no standard answer. It involves both the temperature and the amount of weight that the person prefers. The use of duvets and light cellular blankets that maintain the temperature at a low weight is increasing, but this may in part be due to the greater number of people who have central heating. There are advantages in keeping the weight of bed clothes at the minimum when the patient is suffering from arthritis, but otherwise no firm guidance appears to be appropriate.

There has for many years been an old wives tale that the quality of sleep is influenced by the weather and particularly by changes in the barometric pressure. Studies have demonstrated that there is a measure of truth in these feelings, for both very high and very low barometric pressures are associated with increased sleepiness as determined by e.e.g. recordings. The view that sleep is better during certain periods of the year is explained on the basis of both the barometric pressure and the ambient temperature rather than an effect similar to the hibernation experienced by animals.

Exercise

Athletes who are doing a fairly constant amount of exercise have a greater proportion of deep sleep than others, but there is no consistent finding that sudden strenuous exercise in non-athletes leads to an increase in sleep depth or length. On the other hand when athletes in training reduce their amount of exercise, it does reduce the amount of their deep sleep. In general rather strenuous exercise seems to have its greatest benefit when taken in the afternoon or the early evening rather than in the morning. Levels of exercise that are high enough to give rise to stress late in the evening can produce a higher level of arousal at bedtime and be counter-productive in improving the sleep pattern. Surprisingly, confinement to bed with consequent low levels of activity increases the depth of sleep rather than reduces it. Hence, as with most aspects of sleep hygiene no consistent pattern emerges that will suit even the majority of persons.

From practical experience it is our opinion that some exercise in the fresh air during the hour before retiring is often beneficial, particularly if it is followed by a relaxing hot bath just before bedtime.

It is particularly helpful in those who are studying hard and are tense just before exams. Whether the improvement is due to the exercise *per se* or whether it means that work is abandoned a little earlier and a measure of relaxation occurs is a matter of discussion. Sudden strenuous exercise has no place as a measure to encourage sleep. It leads to aches, pains and stiffness which reduce the possibility of sleep rather than improving it.

Food

There can be few of us who are not aware that a reasonably large lunch, a comfortable chair, an equitable temperature, a darkened room for the demonstration of transparencies and the droning voice of a poor lecturer are ideal for encouraging a quiet doze. Hence we are well aware that food intake does influence sleep. Weight gain, when the food intakes is above the body needs, leads to quiet and long sleep periods, and weight loss, as a result of a low food intake, leads to more fragmented and shorter sleep. Nor is this surprising when it is appreciated that in the natural state, an animal which is losing weight will be in need of high arousal to capture larger amounts of food.

The influence of food is one of the few areas of sleep hygiene in which there is some measure of agreement across the population. Food usually assists sleeping, a factor which has been utilized by at least one food manufacturer who ties the product, taken late at night, with sound sleep. It is very doubtful whether such a relationship exists, but the general factor of an easily digested, warm source of food has considerable merit.

On the other hand it is counter-productive to indulge in heavy meals just before retiring, particularly in those who suffer from indigestion. Moderation is important. Indeed, there is no clear evidence that any particular food is any more beneficial than another, though on theoretical grounds it might be supposed that food which has a high content of L-tryptophan might be indicated. However the levels of tryptophan in even high protein foods is low, and little tryptophan enters the brain due to the competition between that amino acid and others for the carrier mechanism.

Fluid

Too low a fluid intake, with a stressful dry sensation in the mouth is not conducive to good sleep. On the other hand too high an intake of fluid during the evening which inevitably leads to a full bladder during the night is a frequent cause of poor sleeping. This is particularly true of the elderly with diminished bladder volumes. An additional factor is the nature of the fluid. If alcohol, diuresis is encouraged and chronic ingestion will itself lead to a poor sleep pattern. Many other common beverages, e.g. tea, coffee, cola contain significant amounts of caffeine and related xantheine derivatives and these too can reduce the chance of a good sleep pattern. In those with insomnia, the intake of fluid during the later part of the evening should be restricted. When a drink is taken there are merits on it being based upon milk, which due to its calorie content may improve sleep.

Medication

Sleep is made worse by stimulant compounds. Such substances whether they are therapeutic, in food and drink (e.g. caffeine) or involving drug misuse should be avoided when there is a problem of insomnia. Sedative compounds, as their name implies, will usually, for a time at least, improve the sleeping pattern. However, with continued use, tolerance is common and the sleep pattern deteriorates. Both stimulant and sedative compounds can, under particular circumstances give rise to sleep disturbances.

Special Techniques in Sleep Hygiene

Relaxation Techniques

Progessive muscle relaxation is one of the techniques that has been used successfully in problems in getting to sleep. The technique consists of a series of exercises that contract and relax various muscle groups. The individual groups of muscles are contracted serially and then relaxed, and attention is focussed on the sensation of relaxation that can be achieved. Usually each session is devoted to a separate area of the body. The patient is taught to contract groups of muscles of all parts of the body, to identify when moderate levels of tension exist therein and to know how to relax them. The patient continues to practice the techniques in his own home. Instructions, either written or on cassettes, are available to assist them.

The method is easy to teach, and is generally well received by patients who readily identify the procedure as relevant to their problems. Sleep latency is said to be reduced though the results are variable. The best results are found when the sleep disturbance is associated with an anxiety state.

Sleep Conditioning

This technique is based upon the concept that as a result of years of difficulty in sleeping the patient has come to associate the bed with the frustration of insomnia. Even if the original stress that caused the difficulty has disappeared the bedroom triggers the emotions which disturb sleep. Reinforced by the occasional naturally occurring poor night's sleep, such a conditioned insomnia will maintain itself for a long period. Such patients find that most changes in their circumstances of sleep lead to an improvement. Thus such people, unlike normal people, tend to sleep better away from their own homes, even on the first night. In their own home they may be helped by an unusual bed or a different room.

One of the best methods of overcoming a conditioned insomnia is by 'stimulus-control behaviour therapy'. The principle is to associate the normal bedroom stimuli with rapid sleep onset rather than frustration and arousal. To achieve this the patient follows certain rules.

(1) Lie down in bed only when sleepy.

(2) Use the bed only for sleeping; do not read, watch television or eat in bed. Sexual activity is the only exception to this rule. On such occasions follow the sleep instructions afterwards.

(3) If unable to fall asleep within 10 minutes get up and go into another room. Stay up until there is real sleepiness, and then return to the bedroom to sleep. Get out of bed again if sleep does not come easily.

Remember, the goal of this procedure is to associate the bed with falling asleep quickly.

(4) If sleep does not come, repeat step (3) as often as is necessary throughout the night.

(5) Set the alarm and get up at the same time every morning regardless of the amount of sleep during the night. This will help the body acquire a constant sleep–wake rhythm.

(6) Do not nap during the day.

During the first night or so the sleep is normally very poor, but with support and encouragement from the doctor many patients find that they can follow this regime, and the results may be very encouraging.

Biofeedback

Subtle variations in muscle tension can be detected by electrical recordings which convert the energy into auditory signals. By listening to the frequency of clicks from a loudspeaker, the patient can be made aware of muscle tension and learn to control it. This technique of biofeedback is a modification of relaxation therapy. An even more fascinating prospect is to provide patients with a biofeedback of the electrical activity of the brain. A selected pattern is displayed to the patient as an auditory signal which enables him to distinguish the signs of arousal from those of the initial stages of sleep. The patient then learns to relax so that the signs associated with drowsiness are predominant and sleep is encouraged. While these techniques have hardly reached the stage of general clinical use, the experimental results are encouraging.

All the methods of encouraging sleep without the use of sedatives which fall within the category of sleep hygiene must be regarded as highly desirable. Insomnia is a common problem, there are worries about the level of long-term use of sedatives and given training and encouragement, most patients can apply the techniques themselves with great success.

Chapter 5

Hypnotics

Mode of Action

The action of hypnotics can be viewed from two points of view: the brain areas at which the main effects are exerted and the cellular mechanisms involved. As has been explained earlier, the sleep–wakefulness balance results from an inherent rhythmicity of portions of the brain stem which can be modified by external influences. In consequence during wakefulness, the cerebral cortex is stimulated by the reticular activating system via the thalamic nuclei.

Sleep is encouraged by reducing the activity of the reticular activating system, by blocking the impulses it sends to the cortex or by reducing factors which drive the reticular activating system. One of the typical drives for the reticular activating system is a painful stimulus. It is by blocking these influences that analgesics can encourage sleep in those who have been kept awake by chronic pain (Figure 19).

For some hypnotics, the barbiturates, the mode of action is to block transmission of activating impulses both in the reticular activating system and at the cortical level (Figure 20). In consequence there is a general retardation of cortical activity which may explain why the long term administration of the barbiturates may lead to cerebral degeneration.

The reticular activating system is influenced by the 'limbic system' – neuronal loops involving the temporal lobe concerned with emotion. The input of the limbic system is such that when these emotional circuits are at their quietest sleep is encouraged. Conversely, emotional

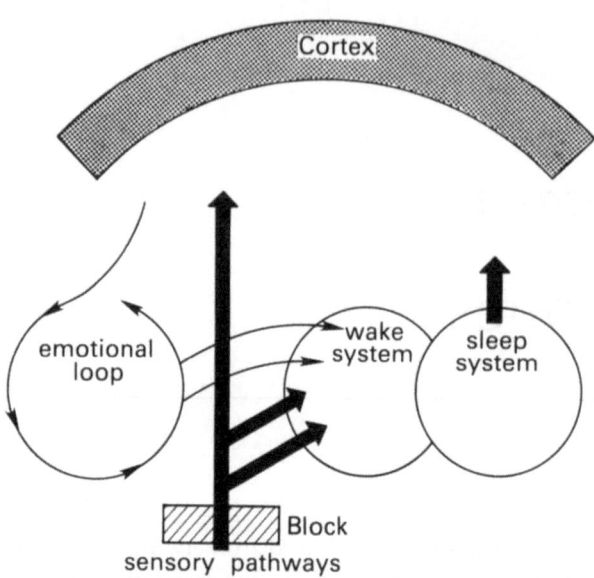

Figure 19. Compare with Figure 12. A schematic representation of the mechanism by which analgesics overcome the insomnia associated with pain by blocking the afferent pain pathways

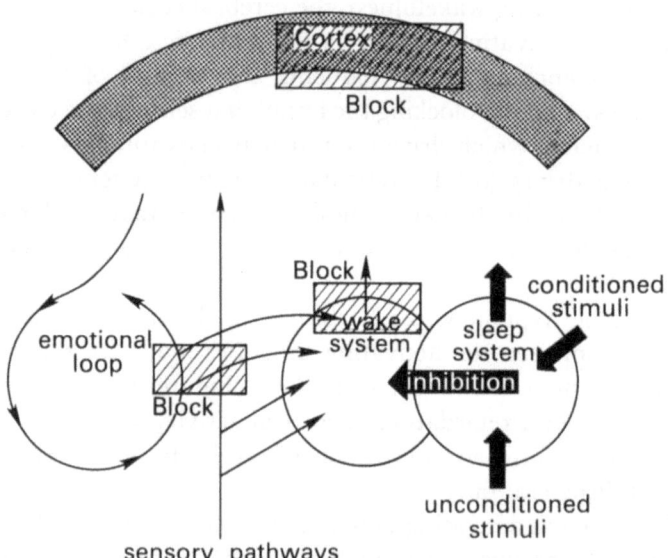

Figure 20. A schematic representation of the widespread effects of barbiturates used at normal hypnotic dosage

upsets encourage an active brain with reduced chance of sleep. Studies over the past decade have suggested that one of the main sites of activity of some hypnotics, particularly the benzodiazepines, is to depress the activity of the limbic system, so reducing the stimulus to the reticular acting system and encouraging relaxation of the cortex for sleep (Figure 21).

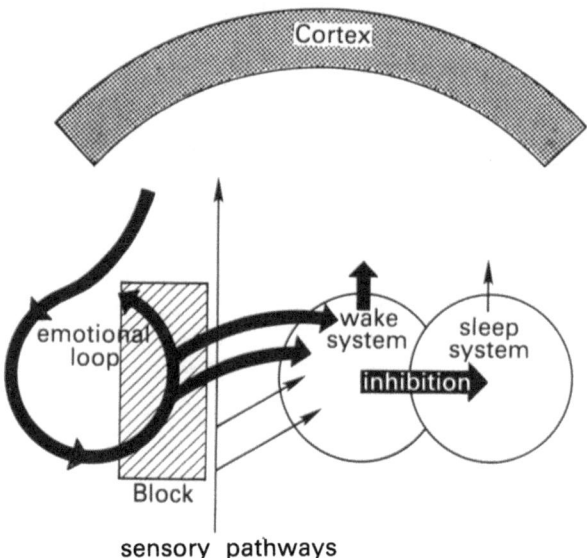

Figure 21. A schematic representation of the effects of therapeutic doses of sedative benzodiazepines. Note the greater specificity of action compared with barbiturates (Figure 20)

All these activities depend upon the transmission of nerve impulses. Such impulses travel along neurones as a wave of ion movement generated electrical pulses along the membrane surfaces. At the synapse the transmission of the impulse depends on a transmitter substance from the axon terminal interacting with specific receptors in the postsynaptic membrane. The current view is that the transmitter substance that encourages sleep is 5-hydroxytryptamine (5-HT, serotonin) while that which reduces the limbic system activity is gamma amino-butyric acid (GABA) (Figure 22).

As a result of the interaction of transmitter and receptor at the postsynaptic membrane, channels that permit the flow of ions open. Ion movement occurs, depending on the relative concentrations of the

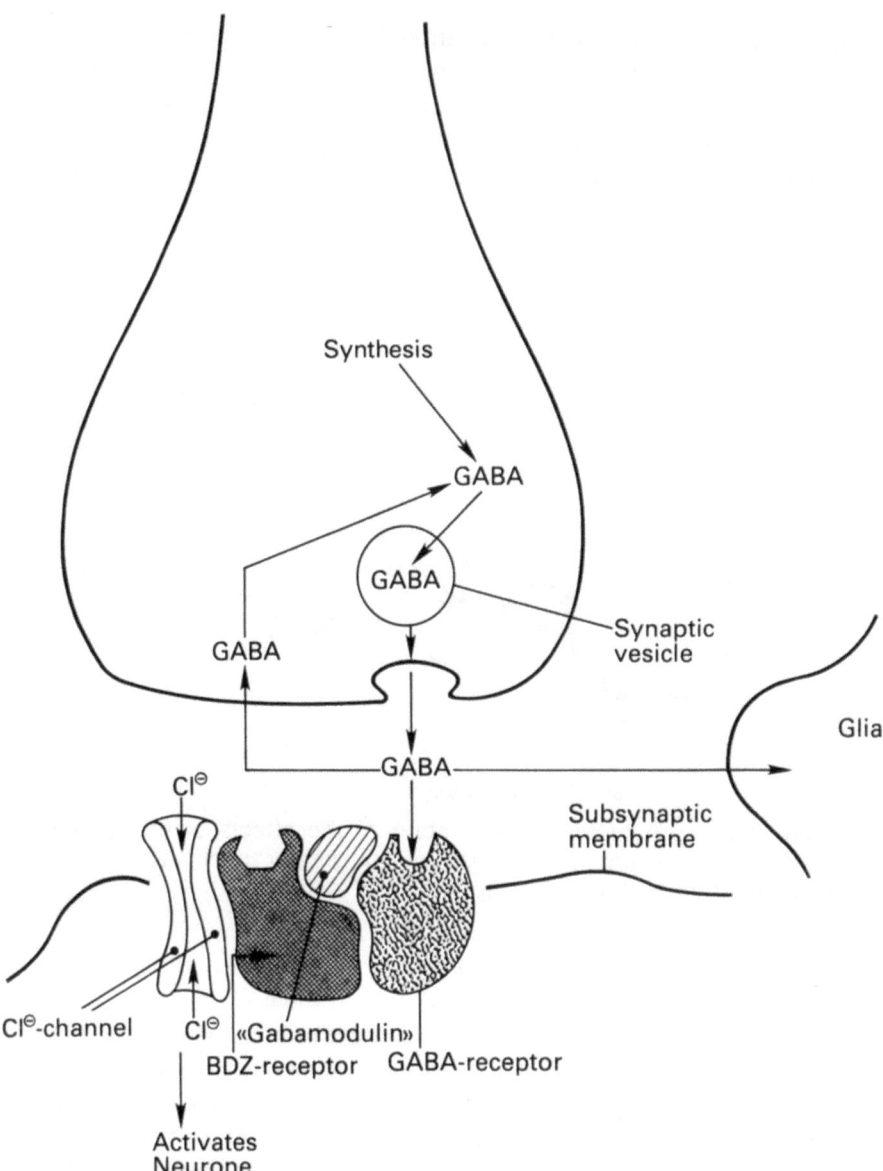

Figure 22. Schematic representation of a gabaminergic neurone influencing the postsynaptic neurone by an effect on the chloride channel. (Modified from Haefely, W. *et al.* (1983). Neuropharmacology of benzodiazepines: synaptic mechanisms and neural basis of action. In Costa, E. (ed.). *The Benzodiazepines: From Molecular Biology to Clinical Practice.* pp. 21–66. [New York: Raven Press]

ions inside and outside the membranes, and the nature of the channels that are open. Sodium flow inwards leads to excitation while chloride flow inwards or potassium flow outwards causes inhibition.

Barbiturates appear to interact with receptors that are very closely related to the chloride channels (Figure 23). This interaction increases

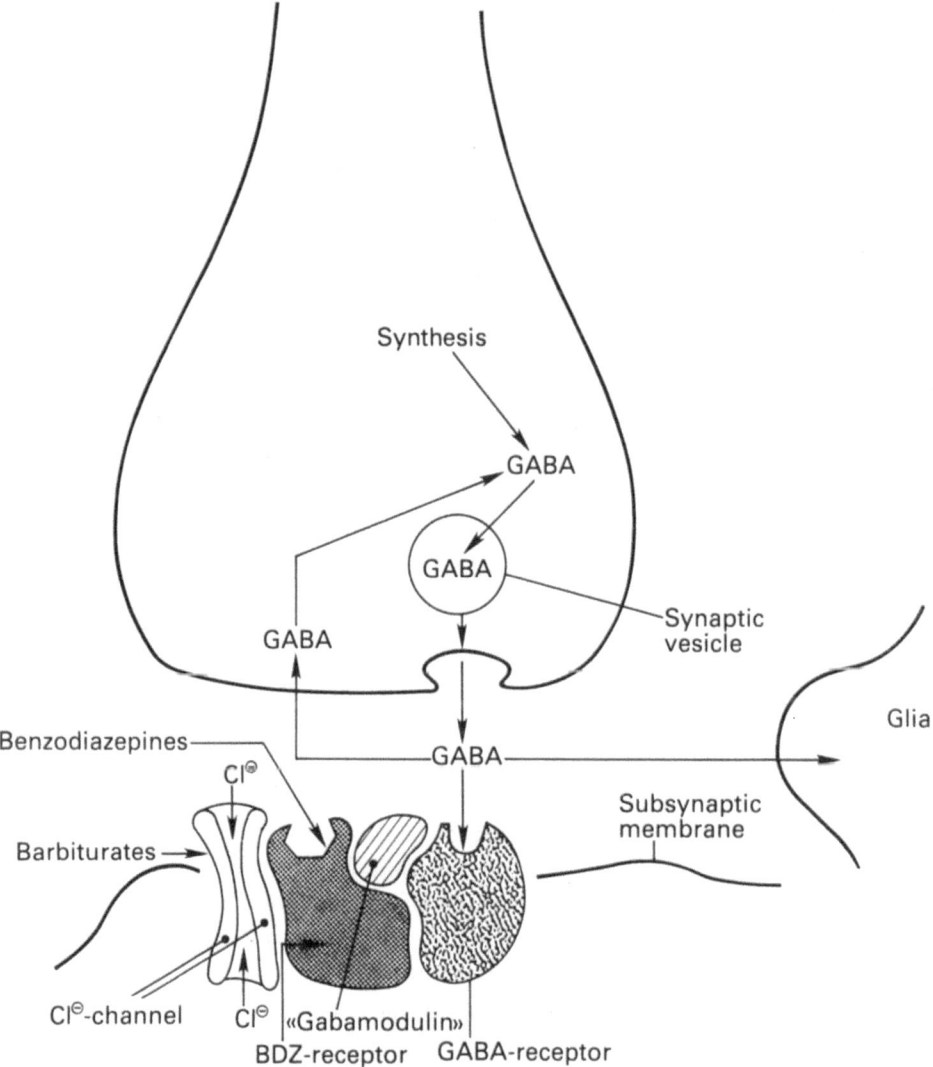

Figure 23. Compare with Figure 22. Representation of postulated mode of action of barbiturates (on chloride channel) and benzo-diazepine on separate (BDZ) receptor in gabaminergic synapse

the length of time that the chloride channel is open, and the resulting movement of the chloride ions leads to inhibition of these cells and sleep. The sedative effect of alcohol may be related to a similar activity, though the site of interaction with the chloride channel may differ from that of the barbiturates.

Within the limbic system the GABA receptor site appears to be linked to chloride channels. When GABA is released from the axon terminal across the synaptic cleft, the chloride channels open and inhibition results. Any substance which binds to the GABA receptor should lead to emotional changes which, depending on the ion effect, might be inhibitory (anxiolytic) or excitatory (anxiety provoking). Although substances are known that demonstrate each of these effects experimentally, no such substances are currently available for use in the clinic.

However, there are other receptors closely related to the GABA receptors to which benzodiazepines appear to attach. They are currently known as the 'benzodiazepine' or 'BDZ' receptors. The benzodiazepines do not themselves open chloride channels or inhibit cells. They activate the GABA receptors, so that when GABA is released the chloride channels open and the cells are inhibited. There is a large group of substances, including the benzodiazepines, which attach themselves to the 'BDZ' receptors (Figure 23). Many of these exert the characteristic effects of anxiolysis, sedation and muscle relaxation. However, some block the sedative effects of the benzodiazepines (antagonists) or provoke anxiety reactions or convulsions. The reason for these contrary effects and for differences in clinical activity, i.e. greater sedative activity, is unclear, but different types of BDZ receptors have been postulated.

These receptors have raised the possibility that there are natural substances (ligands) which can attach to them. Such ligands might be the natural cause of anxiety, the natural relief from it or the body's normal hypnotic; one such hypothetical influence shown on Figures 22 and 23 is 'Gabamodulin'. This is only an interesting concept for no natural ligands have yet been determined. The different modes of action of hypnotics may explain some of their clinical differences – for example the high risk of death with barbiturate overdosage.

Clinical Pharmacology

Hypnotics are extensively prescribed, and practitioners should understand the basis of the clinical use of these drugs. This includes their mode of action, their pharmacokinetics and pharmacodynamics and the implications thereof, their risks and contraindications. Without such understanding, rational therapy in terms of selection, dose, expected and adverse effects is difficult – if not impossible.

Pharmacokinetics

After the ingestion of an hypnotic three processes are involved in the changes which occur in plasma levels. These are absorption from the gut, distribution into the tissues, and elimination. They proceed simultaneously, with each exerting some influence on duration of action. It is for these and for other reasons that using the elimination half-life alone to classify hypnotics into so-called short, medium and long acting is misleading. Absorption and distribution must also be considered.

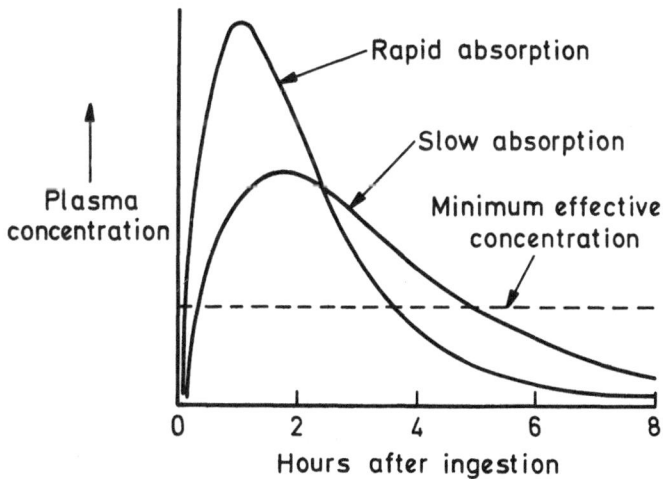

Figure 24. Plasma concentration curve for a substance that is slowly absorbed. The threshold line shows that there may be a prolonged action despite the relatively short elimination half-life

Absorption

With oral ingestion the major determinant of the onset of action of a single dose is absorption (Figure 24). Rapid absorption is usually associated with a quick onset of action, and with slow absorption activity may be attenuated or even absent. When a drug is taken as an hypnotic an adequate rate of absorption is desirable, whereas a slower absorption rate may be preferred if a sustained anxiolytic effect without immediate drowsiness is sought.

Distribution and Elimination

In the body an hypnotic is distributed into the central compartment of blood and highly vascular tissues such as the heart, lung and liver, and a peripheral compartment of lesser vascularity such as voluntary muscle. The brain is a highly vascular organ, and as hypnotics depend for their action on the fact that they cross the blood–brain barrier with ease, the brain is also part of the central compartment. Indeed, brain concentrations follow closely those of the plasma.

When a drug is given intravenously there is for all practical purposes instantaneous mixing with the plasma and the changes in plasma concentration, which are usually assessed by measurement of venous plasma, indicate the parts played by the central and peripheral compartments. However, when a drug is given orally or intramuscularly, plasma concentrations are influenced by absorption, and so there is a growth phase. Since penetration of the blood–brain barrier is good, absorption is the factor which tends to limit the rate of transfer of an hypnotic to its site of action.

The first part of the decay in the plasma concentration relates primarily to distribution into the tissues and so to penetration of the peripheral compartment, while the second part relates to the elimination by metabolism and/or by excretion from the central compartment (Figure 25). Tissue penetration and elimination occur together, and the fall in plasma level reflects the relative dominance of these events at different times.

A drug has an effect as long as its plasma concentration remains above a certain plasma level. If the fall below the threshold is reached during the phase which predominantly represents distribution – and

70

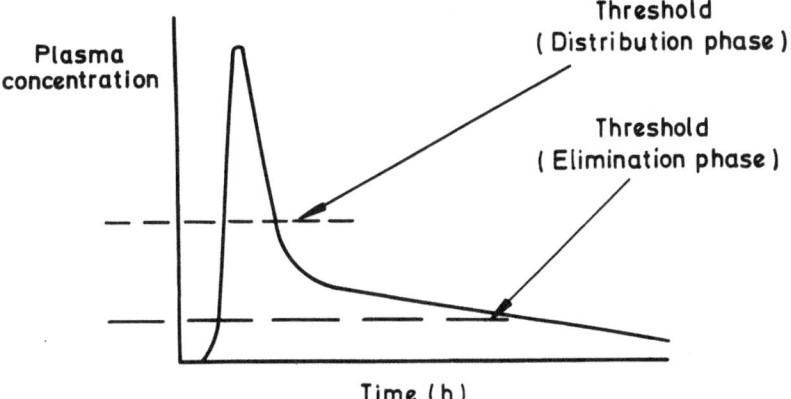

Figure 25. Plasma concentration profile (semi-log plot) of a drug after oral dosing. There is an effect as long as the concentration remains above a certain level. This threshold may be related to the distribution or elimination phases. If it is above the concentration at which the inflexion of the distribution and elimination half-lives occurs then it will be related to distribution and any effect will be of short duration, but if it is below the inflexion then it will be related to elimination and any effect will tend to persist.

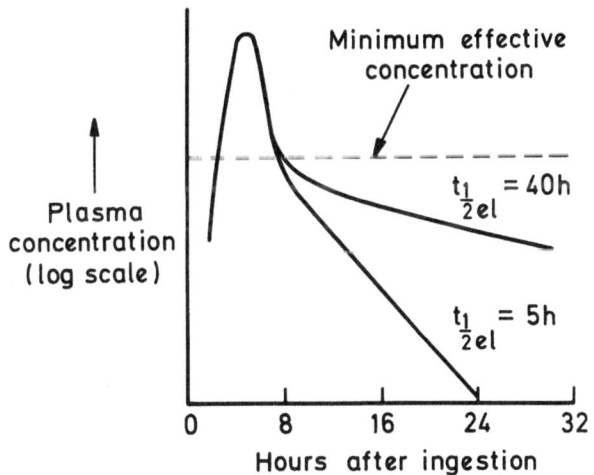

Figure 26. Plasma concentration curve for a substance that is rapidly absorbed, has a rapid tissue distribution and a relatively long elimination half-life (40h). The threshold line shows that there may still be a short duration of action after a single dose despite the long elimination half-life

71

this part of the decay is usually rapid – then the duration of action is likely to be short, but if the threshold is such that it is only reached in the elimination phase then length of action will be much longer – particularly if the elimination rate is slow.

Distribution and elimination influence duration of activity, and so a relatively short duration of action may be attained even though elimination is relatively slow (Figure 26). However, the drug must become available in the blood in relatively high concentrations soon after administration, otherwise the concentration profile will level out and no distinction will be found between the faster distribution and slower elimination phases.

Intermittent Therapy and Accumulation

In the management of sleep difficulties an intermittent type of drug action is often desirable – even when the drug is given daily. The theoretical plasma profile levels of two hypnotics with very different half-lives are illustrated in Figure 27. When given every 24 hours a drug with an elimination half-life of 24 hours will accumulate, but one with a half-life of 6 hours will not, and this ensures an intermittent type of

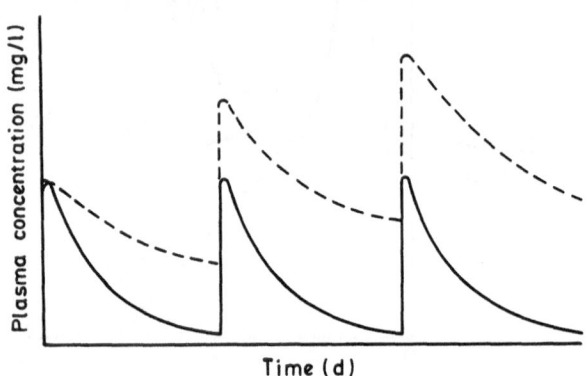

Figure 27. Plasma concentration profiles of two hypnotics with different elimination half-lives (lower curves – 6 h: upper curves – 24 h) given every 24 hours. If the compound with a half-life of 6 h is given daily there will be no accumulation, and there will be an intermittent type of drug action. However, the drug with an elimination half-life of 24 h will accumulate with daily ingestion

action. However, clinical effects may not relate directly to plasma concentration. Tolerance may develop, and this is particularly true for the central nervous system. Indeed, unwanted drowsiness may be experienced only early on in the course of chronic therapy. However, tolerance to the persistent effects of an hypnotic is not a point in favour for its use when there are other suitable hypnotics free of persistent effects.

The rate of accumulation of a drug varies inversely with the elimination half-life, and so the longer the half-life, the slower the rate of accumulation. In general, steady state conditions are reached after an interval of about four to five times the elimination half-life. Drugs with long half-lives accumulate slowly but extensively, whereas accumulation, though completed more rapidly, may not be a factor of clinical significance with drugs with shorter half-lives. Elimination of benzodiazepines occurs either by excretion of the compound itself, or by conjugation to glucuronic acid or after biotransformation. In general compounds that are not metabolized before excretion are those with shorter elimination half-lives.

Pharmacokinetics of Hypnotics in the Elderly

Absorption and distribution show no significant differences from the young adult. Moreover, unless there is severe renal impairment the excretory component of elimination is not affected. Benzodiazepines which do not require significant biotransformation before excretion have a pharmacokinetic profile similar to that seen in the young adult. On the other hand, the metabolism of benzodiazepines is likely to be influenced by the age of the patient. Individuals over 60, even when healthy, may have some impairment of the ability to complete the usual biotransformations, and so the half-lives of long-acting benzodiazepines tend to be prolonged in the elderly as opposed to the young. Age related decrements can vary from slight to a very marked effect depending on the drug and on the sex of the patient. Metabolic impairment in elderly males may be greater than that in elderly females.

Pharmacodynamics

Pharmacodynamics take account of the overall effect of absorption, distribution and elimination, and considers these in relation to inter-actions at the cell interface. To understand the pharmacodynamics of the benzodiazepines it is convenient to consider two representative members, diazepam and lorazepam.

Diazepam is quickly absorbed. A single dose of say 10 mg, has an immediate effect, which persists for a few hours even though its elimination half-life ranges from 14–90 hours. This is due to rapid and extensive distribution into the tissues. However, daily ingestion will lead to accumulation both of the parent compound and of its long-acting metabolite, nordiazepam, which also has a long elimination halflife (30–60 hours). It is therefore important to appreciate that residual sequelae with a single dose of 10 mg diazepam are unlikely, but that a sustained anxiolytic effect will follow daily ingestion.

The pharmacokinetics of lorazepam are different. With lorazepam the distribution phase is less extensive, and so concentrations above the threshold for a particular effect are more likely to be related to the elimination phase with its half-life of 10–20 hours. Though this is much shorter than that of diazepam, a *single* dose of lorazepam may have a more prolonged effect than diazepam due to the absence of a sustained distribution phase. The repeated use of benzodiazepines with half-lives of 16–24 hours or longer, of either the parent compound or of an active metabolite, usually lead to persistent activity and this property is relevant to anxiolytic therapy.

Despite the fact that the interaction of the benzodiazepines and their specific receptors have been studied extensively over the past few years, one aspect that is still far from clear is the time sequence of the interaction. The time sequence determines whether the plasma level represents the effect within the cell or whether the cell interaction persists when the plasma level has fallen, and until this is known we cannot claim a complete understanding of the relations between pharmacokinetics and pharmacodynamics.

Residual Sequelae

Obviously, hypnotics with limited duration of activity are appropriate for those involved in skilled activity and in whom the predominant problem is that of disturbed sleep. However, residual impairment of performance with overnight ingestion may arise and may extend well into the next day – even 'with doses which are within the generally accepted therapeutic range. A variety of tasks have been used to investigate this problem, and there is broad agreement on the relative persistence of the various drugs available. Sequelae are related to dose both in the extent of the decrement at any given time and in the persistence of the impairment, and to the pharmacokinetic profile. Impaired performance is more severe and persists far longer with higher doses of long-acting compounds.

The importance of residual effects depends on both the clinical indication for the use of the benzodiazepine and the occupation of the patient. In those whose insomnia is associated with anxiety, the use of benzodiazepines with next day activity is appropriate. Furthermore, the occasional omissions of doses will not lead to an acute re-appearance of symptomatology. However, accumulation may give rise to unwanted daytime drowsiness and impairment of psychomotor performance. Nevertheless, with low doses impaired performance the next day may be minimal, and this may be offset, at least partially, by tolerance. Further, performance in some patients with anxiety may even be improved by therapy.

For patients with insomnia not associated with anxiety, and where, due to the patient's occupation, impaired performance the next day is not acceptable, a rapidly eliminated benzodiazepine with a duration of action compatible with the clinical sleep problem should be used. Very rapid elimination may be appropriate when the patient experiences difficulty in getting to sleep again having woken up unexpectedly during the night (e.g. a businessman who has crossed time zones) and when the complaint is getting to sleep. A somewhat longer duration is appropriate if repeated awakenings during the night is the main feature. In the latter case with some hypnotics there may be some performance loss the next morning, and the patient should be warned as this may not always be apparent clinically – particularly after the first few days of therapy.

Rebound

Rebound is a common feature of the use of many drugs withdrawn

75

suddenly – particularly after high dosage for a prolonged period. In the case of hypnotics it may take the form of an exaggerated level of insomnia and possibly anxiety. With short-acting benzodiazepines, rebound may occur immediately after withdrawal and may be pronounced, but with longer-acting drugs the reappearance of pre-treatment symptoms of insomnia or anxiety may be delayed for several days and may not even be accentuated.

Though it would appear that rebound insomnia and anxiety may occur after sudden cessation of relatively high doses of benzodiazepines for long periods of time, there is little evidence that they occur if the drugs are used correctly, i.e. in low doses and, in the case of hypnotics, occasionally.

Chapter 6

Selecting an Hypnotic

Nowadays, the sedative of choice is normally one of the benzo-diazepines, though at least one hypnotic of a different chemical structure (zopiclone) is under clinical trial. However, the benzodiazepines are effective, and they are remarkably safe for active compounds. They have a wide margin between effective and toxic doses ensuring that death from accidental or deliberate overdosage is rare. Moreover, organ damage is unusual, and even prolonged administration gives very few problems with negligible evidence of tolerance. The only significant problem is the development of dependence, both physical and psychological. Such dependence occurs with most, if not all, compounds that depress the central nervous system, but current evidence suggests that the risk is low, certainly substantially below that of other commonly used sedatives, such as the barbiturates.

For all these reasons, at least for the immediate future, most patients with insomnia who need an hypnotic are likely to be treated with a benzodiazepine, and the question is how to select the most appropriate one for the patient. The selection depends on:

(1) The nature of the sleep disturbance. Is there delayed sleep onset or multiple awakenings (possibly both), early morning awakenings or insomnia with anxiety?
(2) Does the patient's work or domestic situation require alertness early the next morning?
(3) Are occasional or nightly doses likely to be needed?

There are, however, two exceptions to the general rule that a benzodiazepine is the drug of choice. This concerns insomnia arising from depression and that associated with sleep apnoea. In patients with depression a sedative antidepressant should be given, and in the elderly it should be free of anticholinergic activity. In some cases, for example when there is associated anxiety, the antidepressant can with advantage be given with a benzodiazepine to increase the sedative component. In patients with sleep apnoea, benzodiazepines are always contraindicated.

These considerations are then related to the pharmacological profile of each hypnotic, including that of their metabolites. In broad terms an adequate rate of absorption is needed when there is difficulty in falling asleep, a sufficient duration of action is required for patients who have difficulty in staying asleep, and an effect extending into the next day may be useful for those with daytime anxiety, though there are many other factors to be borne in mind.

Active patients may require an hypnotic free of any early morning residual sequelae, and so a drug which is rapidly absorbed and in which the plasma concentration falls during the sleep period below that of the minimum plasma level for impaired performance is needed. On the other hand, though most, but not all, hypnotics are rapidly absorbed and induce sleep quickly, effectiveness in maintaining sleep varies, and so benzodiazepines which are very rapidly eliminated may not be appropriate if difficulty in staying asleep is the problem.

Currently Available Benzodiazepines

There are now nearly 40 benzodiazepines which are either commercially available in some or many countries, or are under study and likely to be available in the near future. The majority of these produce some measure of sedation and can be useful for insomnia. However, it is reasonable to use only a few, though experience leads to different recommendations. Some authorities, for reasons often difficult to understand, or to relate to clinical pharmacology, or to medicine, centre their recommendation on a particular drug, but we feel that such an approach is unrealistic, and not acceptable to current medical practice. Indeed, there are specific indications for the various hypnotics. We have also included data on some hypnotics still under development, as these may prove to be of particular use in the near future.

The benzodiazepines which should be considered are:

Nitrazepam (Mogadon) was one of the first benzodiazepines and is still used extensively. It is relatively slowly eliminated, and accumulates for the first few days on daily ingestion. Though tolerance to daytime effects may develop it is only really indicated if some degree of daytime sedation is acceptable. Such an hypnotic is of value in patients with nocturnal and early morning awakening when early morning alertness is not essential. The usual dose is 5 mg, and 2.5 mg is recommended for the elderly.

Flurazepam hydrochloride (Dalmane) has an active metabolite which is slowly eliminated. It is a safe hypnotic with high patient acceptability, and has a useful place in the treatment of insomnia, particularly in those in whom nocturnal and early morning awakenings are the major problems. Daytime alertness is impaired by 30 mg overnight, but such impairment is likely to be minimal when the dose is reduced to 15 mg, and this is now the recommended dose in the United Kingdom. However, even at this dose, there is some accumulation on daily ingestion. It is effective for weaning patients from the chronic use of barbiturates.

Flunitrazepam (Rohypnol) is rapidly absorbed and this is a consistent feature of the drug. The plasma concentration falls quickly during the distribution phase, but it is relatively slowly eliminated and so may accumulate to some extent. The recommended dose range is 0.5–1.0 mg. The lower dose is free of residual effects the next day, but residual effects have been reported with 1.0 mg, though they are minimal. Further studies are needed to decide the place of this drug in therapy, but the rapid onset of effect suggests that the lower dose (0.5 mg) may be useful for those who have difficulty in falling asleep and need an hypnotic free of residual effects the next day.

Diazepam (Valium) is a useful hypnotic even though it was not introduced as such into clinical practice. The plasma concentration falls quickly during the distribution phase and so does not lead to impaired performance during the morning after a single dose of 10 mg overnight. Unfortunately, due to slow elimination of the parent compound and one of its primary metabolites, nordiazepam, daily ingestion leads to accumulation. To avoid residual effects, diazepam should be taken not more frequently than once within a 48-hour period and not more than twice in 7 days. In this way it is a useful drug for those involved in skilled activity.

Diazepam used nightly is more appropriate for the patient whose insomnia is associated with daytime anxiety. This is due to accumulation of the parent compound, and its long-acting metabolite. Tolerance to adverse daytime effects is likely to occur within a few days, and with 10 mg nightly such effects may not be a great problem.

Potassium clorazepate (*Tranxene*) is indicated for the relief of insomnia with daytime anxiety. It is hydrolysed in the gut, and absorbed as nordiazepam. A single night-time dose of 15 mg gives effective sleep and relief from anxiety the next day, and in this dose there is little, if any, impaired performance. The daytime anxiolytic effect is evident with the first dose, and in the elderly a lower dose of 7.5 mg is recommended.

Temazepam (*Normison: Euhypnos*) is the other primary metabolite of diazepam. It has a significant distribution phase, an elimination half-life around 10 hours and freedom from long-acting metabolites. There are no residual sequelae the next day after the overnight ingestion of 20 mg, although with a higher dose (30 mg) they may appear. The usual dose is 20 mg, and it is particularly useful for individuals carrying out skilled work, and for shiftworkers. In most countries, including the United Kingdom, it is available in a formulation which is quickly absorbed, and so is suitable for patients who have difficulty in falling asleep.

Oxazepam (*Serenid-D*) is the final metabolite of the breakdown of diazepam. It is slowly absorbed, there is no distinct distribution phase, and the elimination half-life is around 10 hours. There are no residual sequelae after the overnight ingestion of 15–30 mg, but with higher doses (45 mg) they appear. The usual dose overnight is 15–30 mg, and, although it may not be indicated for difficulties in falling asleep, it is very useful for patients who have difficulty in staying asleep, and yet must be alert early the next morning.

Lormetazepam is closely related to temazepam. It has a significant distribution phase, an elimination half-life of about 10 hours and no long-lasting metabolities. The dose range of 0.5–1.0 mg is free of residual effects the next day. With one formulation (Noctamid) the peak plasma level is reached relatively slowly, and so it may not be so useful for patients with difficulty in falling asleep. It is necessary to give a higher dose (2.0 mg) to quicken sleep onset in the chronic insomniac, but unfortunately this dose leads to residual effects the next day. Another formulation (Loramet) with faster absorption is available, and this should be useful for patients with difficulty in falling asleep.

Triazolam (*Halcion*) is one of the triazolodiazepines, and is rapidly eliminated with a half-life around 3 hours without long-acting metabolites. Doses up to 0.25 mg are free of residual effects, and so it is useful for the management of transient insomnia when early morning sedation must be avoided. The rapidity of elimination may not make it ideal for the relief of nocturnal and early morning awakenings. The dose range is 0.125–0.25 mg. The lower dose should be tried initially in all patients, and is, in any case, indicated for the elderly.

Brotizolam (*Lendormin*) is undergoing clinical trials. It is a triazolo-diazepine without long-acting metabolites, but has a longer elimination half-life than triazolam – around 4.5 hours. Brotizolam is without residual effects at doses up to 0.25 mg, and so is useful for the management of transient insomnia when early morning sedation must be avoided. Furthermore, although it is rapidly eliminated, elimination is not so fast as other such drugs, and so it may provide relief from nocturnal and early morning awakenings without residual effects the next day. The dose range is 0.125–0.25 mg with the lower dose for the elderly. A higher dose range of 0.25–0.5 mg may prove useful as an hypnotic overnight preceding surgery.

Midazolam (*Dormicum*) is undergoing clinical trials. It is an imidazobenzodiazepine with an elimination half-life around 2 hours. It will be useful for difficulties in falling asleep, but the rapidity of elimination is unlikely to provide relief from nocturnal and early morning awakenings. Because of its particularly short duration of action it would be an appropriate hypnotic if one was needed during the night. It may prove useful for shiftworkers who have relatively short rest periods. The anticipated dose in adults, including the elderly, is 15 mg.

Table 1 Clinical indications for hypnotics based on pharmacokinetic and pharmacodynamic data (Indications are based on stated dose or dose range)

| | 1,4-Benzodiazepines | | | | | | | | | Triazolo-diazepines | | Imidazobenzo-diazepine | |
	Nitrazepam (Mogadon)	Flurazepam (Dalmane)	Flunitrazepam (Rohypnol)	Diazepam (Valium)	Potassium Clorazepate (Tranxene)	Temazepam (Normison)	Oxazepam (Serenid-D)	Lormetazepam (Noctamid)	Lormetazepam (Loramet)	Triazolam (Halcion)	Brotizolam (Lendormin)	Midazolam (Dormicum)	
Absorption	A	A	Rapid	A	A	A	Slow	Slow	A	A	A	A	A = Adequate rate of absorption. Relatively slowly absorbed drugs may not be useful for difficulties in falling asleep unless taken some time before bed.
Elimination half-life (h)	18–30	24–48	15–30*	14–90*	30–60	4–10*	6–24	c10	c10*	c3	c4.5	c2	* Adequate rate of absorption together with sustained fall in plasma level during distribution phase. This reduces likelihood of residual effects in drugs with relatively slow elimination, e.g. diazepam and flunitrazepam – at least with single doses.
Accumulates with daily ingestion	Yes	Yes	Yes	Yes	Yes	No	Probably not	No	No	No	No	No	

CLINICAL INDICATIONS

Dose (mg) for healthy adults	5	15	0.5–1.0	5–10	15	10–20	15–30	0.5–1.0	0.5–1.0	0.125–0.25	0.125–0.25	15
Initial dose (mg) in elderly	2.5	15	0.5	2.5	7.5	10	15	0.5	0.5	0.125	0.125	**
Initiates sleep without morning or daytime effects				(Yes)†		Yes		Yes	Yes	Yes	Yes	Yes
Sustains sleep, but with possibility of morning and/or daytime effects	Yes	Yes	Yes‡									
Sustains sleep, but without morning or daytime effects							Yes§	Yes§			Yes	
Insomnia with anxiety				Yes	Yes							

Doses relate to general practice and may not correspond to manufacturers' recommendation. Higher doses may be used for in-patients or in psychiatric practice.

** Not yet determined

† Single dose of diazepam is free of residual effects due to sustained distribution phase, but repeated ingestion leads to daytime anxiolytic effect related to slow elimination.

‡ Residual effects are not observed with lower dose of flunitrazepam (0.5 mg) due to rapid fall in plasma level during distribution phase.

§ Slowly absorbed and unlikely to be useful for difficulties in falling asleep.

Table 2 *Suggested uses for specific benzodiazepines*

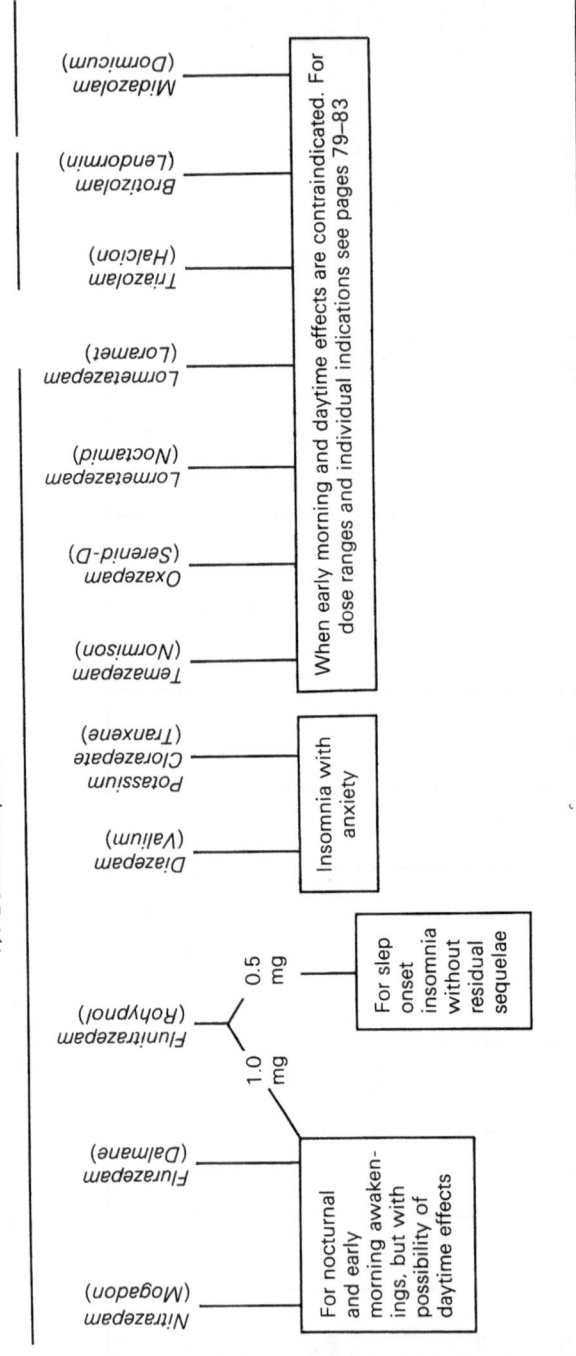

How to Choose an Hypnotic

Tables 1 and 2 are intended to help the practitioner place these benzodiazepines in their clinical perspective. It is unfortunate that clinical trials, which involve comparisons between drugs, do not provide useful information on the relative merits of these drugs. This is because such trials are more than often designed to maximize effectiveness or minimize adverse effects of one of the compounds. This has led to much confusion, not least among those who carry out such trials, though interesting advertisements. Further, they provide information relevant to groups, whereas the practitioner is concerned with the individual patient.

To avoid the invidious and rather pointless assessment of much of the published work on clinical trials with hypnotics, we have considered in Table 1 properties of the benzodiazepines which we believe are of clinical relevance, namely adequate absorption, duration of action, daytime anxiolytic effect, accumulation on repeated ingestion, etc. The practitioner may find that such information is more likely to lead to the correct choice of an hypnotic for the individual patient than reports which highlight the advantages of a particular drug. Table 2 suggests the hypnotics most likely to be appropriate for a specific problem, although it is emphasized, once more, that it is the experience of the practitioner and acceptability by the patient which are the decisive factors in the appropriate use of hypnotics.

Chapter 7

Management of Insomnia

In the management of insomnia the practitioner should keep two factors in mind. Insomnia is a symptom not a diagnosis in its own right, and patients who complain of insomnia are really suffering from 'sleep dissatisfaction'. Secondly, there is no clear understanding of ideal sleep except that which is acceptable to the patient.

Whatever the pressures on time within the consultation, an unravelling of the patient's symptoms and an attempt to determine the cause of the insomnia together with a sympathetic approach, an explanation of the disorder and reassurance are essential. The doctor who reaches immediately for the prescription pad as the patient mentions insomnia is doing neither the patient nor himself a service. He is creating problems for the future.

The diagnosis of any disorder leading to therapy follows a natural path. For most doctors, experience allows the steps in decision making to be completed rapidly and accurately. This is particularly true if the patient is well known to the doctor, who is therefore aware of the background history and personality. However, it is useful to have a decision path available, at least as an *aide memoire*, especially for the puzzling case. A structured approach to the diagnosis of insomnia leading to appropriate management is useful, and a few specific questions are indispensable.

Management Decision Tree

Is the patient tired during the day?   No – see below
                                             Yes – see page 89

This is unlikely to be a case of true insomnia, as such patients are tired during the day. This is probably an example of reduced sleep need.

Action. Explain the varying sleep needs of the individual, and that, if they are not tired during the day, their own needs are being met at their current sleep levels. Give advice about how the patient might occupy, with benefit, the additional hours available to them.

Has sleep hygiene been tried?   No – see below
Yes – see page 90

Action. When there is doubt whether adequate attempts have been made to overcome the problem by sleep hygiene, it is important to give these methods a thorough try. Details are given in Chapter 4.

If such methods are to be tried, the doctor must appreciate that the chance of success is increased if:

(1) The doctor is willing to devote adequate time, support and encouragement.

(2) Empathy exists between the doctor and the patient. This means that the patient needs to be reassured that help will be forthcoming. It may be necessary to demonstrate this care initially in a practical fashion by the provision of one to two good night's sleep with the use of a sedative. Further doses may be necessary at intervals of about a week – each of only one or two days duration – to ensure continued willingness to persist. The aim must be to re-establish a natural pattern of sleep. Persistence and support are essential, but success justifies the effort involved.

Is the onset of insomnia recent?   Yes – see below.

No – see page 91

This is a patient with an acute and possibly transient insomnia, and it is, perhaps, the most appropriate indication for the use of hypnotics. *Action*. Determine the cause of the recent insomnia and treat appropriately.

(1) Change in time zone. Possible need for a sedative over 1–2 days.

(2) Change of sleeping habits, e.g. strange surroundings, shiftwork, hospitalization. Explain the mechanism of the disturbance and give reassurance. It may be necessary to give an hypnotic for a few nights if there is persistent difficulty with reduced daytime functioning.

(3) Bereavement – explain the normal time course of a bereavement and give reassurance. Only give hypnotics if there is severe insomnia leading to daytime distress.

(4) Acute anxiety – deal with the anxiety, trying to remove the causes if possible. Treating the anxiety will also treat the insomnia.

(5) Cardio–respiratory distress. Treat the cause. Since there is likely to be some measure of respiratory failure avoid the use of all sedatives if possible. If this is not possible give very small doses of a benzo-diazepine.

Is there a pharmacological reason?   Yes – see below
No – see page 92

Action. Consider the following possibilities:

(1) Caffeine. Ask about intake of coffee, tea and cola. If these appear to be high, suggest reduction.

(2) Alcohol. Chronic alcoholism can be a cause of poor sleep and requires management of the underlying condition.

(3) Appetite suppressants. Though appetite suppressants are less used than previously, it is important to check that the patient is not using or abusing them. If this is the case deal with the situation by advising earlier use during the day – if essential.

(4) Chronic use of hypnotics – this is a major cause of insomnia and is dealt with on page 97.

Is there a physical cause?   Yes – see below.
                                 No – see page 93.

Action. The management will depend upon the cause:

(1) Pain – define the cause and treat appropriately. Analgesics and sedatives may be needed if cure of the pain is not possible.

(2) Arthritis – this is a special problem, see page 103.

(3) Nocturia – give advice on fluid intake. Consider need for prostatic surgery.

(4) Heart failure – appropriate therapy for the cause. Try to avoid sedatives.

(5) Respiratory failure – treat the cause, and avoid sedatives which will further depress respiratory function.

(6) Gastrointestinal disorders – pain during the night may disturb sleep. If there is a peptic ulcer give antacid just before retiring. If reflux, consider sleep in upright position.

(7) Renal failure. The sleep of patients with renal failure is usually short and disturbed. Dialysis normally improves the insomnia as it improves the general clinical state.

(8) Thyrotoxicosis. One of the manifestations is reduced and fragmented sleep which recovers slowly after surgical cure. Reassurance may be all that is necessary.

Is there a history of intermittent snoring?   Yes – see below.
No – see page 94.

Action. A history of poor sleep, excessive daytime sleepiness and inter-mittent heavy snoring, particularly in patients over the age of 40 and who are overweight, is suggestive of the sleep apnoea syndrome (see page 104).

If this disorder is suspected neurological opinion is indicated. Do not prescribe sedatives.

Are there signs of organic brain syndrome? Yes – see below.

No – see page 95.

Action. Patients with the organic brain syndrome of whatever cause tend to have a very disturbed sleep pattern and frequently experience reversal of the sleep–wakefulness rhythm. This is also true of senile dementia. The problem with such patients, particularly those who are most disturbed, is that their response to sedatives may be irregular, and paroxysmal rage reactions and further disturbance of behaviour are not unusual. All sedatives must be administered with caution.

In such patients phenothiazines may produce effective sedation. The doctor must also consider whether the patient should remain at home or whether institutional care is desirable.

Is there anxiety?   Yes – see below.

No – see page 96.

Action. Anxiety states, either acute or chronic, are a common cause of insomnia. In such patients the insomnia, though it may be the presenting complaint, is secondary to anxiety and it is the latter that requires treatment. When there is a history of chronic insomnia the doctor should ascertain whether there is an anxiety state.

The overall treatment of an anxiety state lies outside the brief for this book. The range of treatment will include psychotherapy, modification of the environment and behavioural therapy. For such patients it *may* be useful to give a short course of an appropriate benzodiazepine, but it is important to stress that such a course should be used to cover only the period of distress and that anxiolytics are not the ultimate answer.

The sedative should be discontinued within 3 months at the very longest to avoid dependence. Further, it should only be given if there is clear evidence that the patient cannot cope without some medication. The decision to start on therapy which may prove to be long term should be very carefully thought through, and fully discussed with an informed patient.

Are there signs of depressive illness?   Yes – see below.
                                             No – see page 97.

Action. Another common cause of chronic insomnia is depression. Classically there is early morning awakening, but other types of insomnia may be seen (page 104). Other signs of depression should be sought in the diagnosis, such as loss of appetite, sadness or crying, early morning apathy, feelings of loneliness, guilt feelings, hopelessness and positive suicidal tendencies. Since such patients are liable to make suicide attempts and are likely to receive some form of sedative it is vital to detect the presence of this disorder. Depression is often masked under an apparent anxiety.

When there is any indication that depression may be present it is vital to treat with an appropriate antidepressant. In the majority of such patients a sedative antidepressant is the most effective. The administration of a sedative antidepressant with a benzodiazepine can often be of benefit.

Has the patient received sedatives
for a prolonged period?   Yes – see below.
                                             No – see page 98.

Action. The action to be taken with patients who have been receiving sedatives for a long period will depend upon the circumstances.

(1) The patient who received hypnotics some years ago for a short term problem; they were never stopped, but the dose has remained constant over the years. Such patients should be withdrawn with support over the critical period. The lack of need for sedatives and their dangers should be explained. The possibility of rebound insomnia must not be hidden, but it should be made clear that it is a brief difficulty. Withdraw the dose gradually, and ensure that the patient receives adequate reassurance and support.

(2) The elderly patient who has been on low doses for a very long time. The kindest action is to continue with the low doses, but ensure that the amount available at any one time is small. This avoids the risk of accidental or deliberate overdose.

(3) The patient with a rising dose level. There is presumptive evidence of dependence. Careful withdrawal is highly desirable, and this is dealt with on page 107.

(4) The patient with chronic insomnia or sleep pathology. See page 105.

Is there an inherent and persistent
difficulty with sleep?   If the patient has not fitted into any of
the previous categories then the answer
is almost certainly yes.

Action. Investigation is usually necessary. It is important to exclude
disorders of the sleep–wakefulness rhythm and sleep pathology such as
narcolepsy, nocturnal myoclonus and sleep apnoea.

In general persistent difficulties with sleep do not respond well to
therapy and their management is difficult. Nevertheless, where a
specific diagnosis can be made appropriate therapy should be tried.
However, while it is necessary to exclude these disorders, the majority
of patients who complain of chronic insomnia are suffering from a
chronic or recurring anxiety state. The difficult problem of the
management of this group is considered in detail on page 105.

Specific Sleep Problems

Transient Insomnia

In patients who complain of difficulty in sleeping it is important to establish the cause (see page 89 *et seq*) and to treat that cause, and not to suppress the symptom. In the majority of patients with transient insomnia, the principles of sleep hygiene with medical support and encouragement will lead to considerable improvement. However, effective support of a patient tired out and distressed by several nights of poor sleep may require the doctor to show his understanding by the prescription of hypnotics for a few days to relieve the distress. Transient insomnia is an important indication for the use of hypnotics, and their use must be approached in a logical manner.

First of all it is helpful to determine whether there is a problem of falling asleep or staying asleep (frequent nocturnal awakening) and also whether there is a significant anxiety element. An hypnotic with an appropriate pharmacological profile should then be chosen. An adequate rate of absorption is needed when there is difficulty in falling asleep, a sufficient duration of action is required for patients who have difficulty in staying asleep and an effect extending into the next day may be useful for those with daytime anxiety.

Patients complaining of transient insomnia will include many active individuals, and so hypnotics free of any early morning residual sequelae may be indicated. For active individuals a drug which is rapidly absorbed and in which the plasma concentration falls during the sleep period below that of the minimum plasma level for impaired performance is needed.

Though most, but not all, hypnotics are rapidly absorbed and induce sleep quickly their effectiveness in maintaining sleep varies. Members of the series which are relatively slowly eliminated may be particularly useful in maintaining sleep. They have proven clinical effectiveness with generally acceptable levels of morning retardation unless critical work is to be undertaken. Such benzodiazepines are described on page 78 *et seq*. However, they are likely to accumulate with daily ingestion.

In the management of insomnia associated with daytime anxiety an anxiolytic effect persisting into the next day may be particularly useful. Several such sedative and anxiolytic benzodiazepines are

99

available. They are useful hypnotics and their daytime anxiolytic effect may be accompanied by little, if any, impairment of performance.

In the management of transient insomnia the lowest dose of an hypnotic should always be used initially, and this dose repeated as infrequently as possible. In general terms hypnotics should not be prescribed initially for more than about 2 weeks, and even for this relatively short period of time they need not be given every night. The patient can be allowed to use his own discretion, and ingestion can be limited to the nights of the week preceding a working day. Alternatively, they can be taken every other night, so that the patient can recall the good sleep of the previous night or look forward to a good sleep during the coming night!

If low doses are used infrequently and for the shortest period of time, many of the problems which may arise in using hypnotics will be avoided. Rebound insomnia, in which sleep is more disturbed on withdrawal of an hypnotic than before treatment, is probably due to daily ingestion of unnecessarily high doses, while dependency arises with regular use over several months. Hypnotics must be given judiciously, and the patient must realize before treatment commences that their use is a temporary expedient.

Bereavement

One specific problem of transient insomnia that practitioners frequently face is that associated with grief after bereavement. The concept of treatment is to support the patient and to provide a brief respite from distress without interfering with the grief reaction. Sedatives may be given for 2–3 days depending upon the patient's needs. It has been suggested that as depression is one of the features of the grief reaction, then an anti-depressant is the drug of choice. We do not accept this view. Anti-depressants may not exert a beneficial effect on depression until about 10–14 days after the start of therapy, and so their usefulness in the short term is only that of their sedative component, and for this reason the benzodiazepines are better and safer.

In most bereaved people, unless there is an overriding need to avoid even early morning sedation, there are merits in using a compound which will provide some anxiolytic effect during the day. In those people in whom it is essential to ensure that there is no risk of sedation the next day there are merits in considering their use only during the weekend.

Middle Aged

Sleep in the middle aged tends to be less restful than in the young, and experience shows that it is in this period of life that many patients receive their first prescription, and so the possibility of chronic therapy arises. In such patients it is particularly important that every effort is made to establish a normal pattern of sleep by other means before more than a few days of an hypnotic is prescribed. In the middle aged, the dose required to produce a therapeutic effect may need to be somewhat higher than in either the young or the elderly, although the same principles of drug selection apply as in the treatment of transient insomnia at any age.

The Elderly

Recent studies have indicated that a substantial number of the elderly, both those who live in their own homes and those who live in institutions, are receiving nightly doses of sedatives. How far this represents the after-effects of too high a level of prescription in middle age and how far it represents a true need for sedatives in the elderly must be a matter of dispute. Certainly, the doses given to the elderly tend to be too high and many started their therapy over a decade ago. As has been explained previously (page 23) the need for sleep in the elderly is reduced, and some prescriptions clearly result from the failure to recognize this.

However, when there is a need for sedatives in the elderly, there are specific factors to be borne in mind. The elderly easily become confused, particularly if there is some measure of dementia. Sedatives increase that confusion, and the patient should receive the initial dose when there is somebody else in the house. One particular problem is insomnia associated with senile dementia. Sleep reversal with night time confusion and activity is common, and there is no really effective and safe hypnotic for these patients. Barbiturates are totally contra-indicated, and current opinion favours either a sedative phenothiazine or a low dose of a rapidly eliminated benzodiazepine.

While benzodiazepine sedatives are the drugs of choice, the elderly show an increased response to the usual dose. The initial dose should be lower than usual until the response is determined. Surveillance should be close as intellectual impairment and postural instability are serious hazards. The standard principles of the lowest effective dose for the shortest period of time apply equally in the elderly as they do in the young.

There remains the vexed question of the management of elderly patients who have received sedatives at low constant dosage for many years. We believe that the kindest approach is to continue therapy if it is not possible to persuade the patient that they are no longer necessary. The quantity of sedatives available at any one time should be restricted so that accidental overdosage is avoided.

Shiftwork

Shiftwork is a fact of life in many types of work, and the exact arrangement of shifts should be designed to minimize the conflicting aims of maintaining the normal circadian rhythm and the social needs of the worker. However, even with good designs, sleep disturbance is inevitable. Shiftworkers usually experience more difficulty in main-taining sleep than in falling asleep. Hence the occasional use of a not too rapidly eliminated benzodiazepine is appropriate.

Transmeridian Flight

Any substantial journey east or west across time zones creates a problem, and the most useful approach is to travel at the right time of day and arrive in time for bed. An hypnotic for sleep during a flight is only likely to be of help during long journeys when there is the opportunity to use the recently improved sleeping facilities. The rest period is unlikely to provide sleep of more than 5 hours' duration, and clearly a low dose of a relatively short-acting hypnotic is indicated.

However, even if sleep loss is minimized during the journey there is a need to adapt to the new environment. Adjustment needs time and some difficulty with sleep for a couple of days is likely. A period of recuperation is needed before working and making decisions. The length should be related to several factors, including the duration of the journey, the number of time zones crossed and the times of departure and arrival. For difficulty with sleep on arrival an hypnotic with a reasonable duration of activity is required. Only 1 or 2 days of a sedative is needed as adaptation will then proceed without the need for further medication.

Pain

Pain itself is only a symptom of another disorder and appropriate treatment is essential. The drugs used for pain vary from aspirin and paracetamol or combinations containing these substances, through non-steroidal anti-inflammatory agents for arthritic disorders, to narcotics for those with an intractable condition. If the relief of the pain is adequate, insomnia usually clears, and a sedative is rarely needed. A small proportion will still suffer from insomnia – perhaps as a result of worry about their present ills or future prospects. In such patients it is useful to give a benzodiazepine with a residual anxiolytic effect since the underlying cause is frequently an anxiety state.

One specific cause of pain is terminal cancer. In such patients it is essential to ensure that there is adequate pain relief, but that daytime sedation is kept to a minimum. There is no place for high doses of hypnotics – adequate doses of narcotics should be prescribed as necessary. One of the best sedatives for such patients is a phenothiazine, and of these chlorpromazine is still preferred by many. Not only is it a reasonable sedative, but it is anti-emetic and potentiates the pain relieving action of the narcotic.

Arthritis

Another common cause of pain sufficient to produce insomnia is arthritis. The pain may be treated with some form of salicylate, one of the non-steroidal anti-inflammatory agents or a steroid. However, even with effective relief, sleep may still be disturbed as a result of stiffness which tends to increase during the night. In such cases there are merits in using a benzodiazepine with muscle relaxant activity.

Other Medical Causes of Insomnia

Insomnia may be a feature of many medical disorders and usually disappears as the condition improves. Insomnia associated with cardio–respiratory disorders with some measure of anoxaemia responds rapidly to relief of the circulatory failure, disturbed sleep from peptic ulceration may be treated with an antacid before retiring, and disturbance due to gastric reflux by sleeping more upright and by the use of sodium alginate.   ·

In some patients there may be a persistent insomnia related to anxiety about their condition. In such patients one of the anxiolytic benzodiazepines may be used with benefit. Particular care is necessary in cardio–respiratory or renal failure and the dose of any sedative should be small.

Sleep Apnoea

There is no satisfactory management that will help all patients with this troublesome disorder. Weight reduction, and some means by which the patient sleeps on his side should be tried. The usefulness of clomipramine is being investigated. If there is an obstruction which cannot be corrected a tracheostomy opened during sleep may be necessary. Alcohol and benzodiazepines, both of which make the condition worse, must be avoided.

Depression

The insomnia of depression occurs classically in the form of early awakening and is relieved as the depression lifts. Benzodiazepines should not be used alone in the treatment of this type of insomnia as the relief of an anxiety component may accentuate the depression. Therapy should involve the use of an antidepressant, and if insomnia is an important symptom then it is useful to use a sedative antidepressant or to administer a benzodiazepine at the same time.

Chronic Insomnia

These patients still represent a major problem in management. They can be considered under various categories depending on their history.

(1) Elderly patients who receive low doses of barbiturates are encountered in most practices. They received a prescription for a barbiturate many years ago and are still taking the tablets because there was no firm attempt to stop the medication. There is no evidence of mental deterioration or dose escalation due to tolerance. In some of these patients it will be relatively easy to stop treatment if time is spent over the explanation. In others there will be difficulty over accepting the advice, and in these we suggest that the best practical procedure is to give no more than 1 month's low dose supply at one time, to insist that the prescription records are checked regularly to ensure that the patient is returning at intervals which accord with the dose, and to warn that the tablets should be kept safely.

(2) Patients who receive increasing doses of barbiturates, and who show tolerance with rising dose levels should be weaned off in the manner described on page 107. This requires substitution by a long-acting benzodiazepine of adequate strength to ensure that the patient sleeps well and accepts the substitution. The dose is then withdrawn in a stepwise manner, and the patient given adequate support. With such support most patients will accept the plan and succeed.

(3) Patients suffering from a severe and intractable chronic or relapsing anxiety state with insomnia may find it impossible to cope unless they are supported with drug therapy. There is still doubt about the proportion of patients who fall into this category, but all those who work in the field can testify to the existence of such patients. Care is necessary not to overdiagnose the condition and this may lead to excessive and prolonged therapy with the risk of psychological and physical dependence. The diagnosis should only be made when it has been shown that the patient will not respond to any other form of therapy.

These patients may experience considerable distress. The doctor has a duty to relieve this distress, unless there is clear evidence that by so doing the condition itself is aggravated or that the risk of therapy is greater than the original problem. If doctors refuse to treat such patients there is a distinct risk that the patient will resort to other forms of relief that are far more dangerous and less effective, e.g. alcohol.

In such patients there is no reason for not treating the anxiety over the long term, and this can best be done with one of the less rapidly eliminated benzodiazepines as they are sufficiently safe for long term use to be justified.

The main danger lies in the possible development of dependence particularly as many such patients show evidence of a personality that favours the development of dependence. However, even in such patients, e.g. alcohol abusers, a substantial proportion can be treated for prolonged periods without problems so that the risk should not be overstated.

The need for the continuation of therapy should be assessed during the second month of therapy. In the meantime alternative methods should be tried. If possible the situation and the potential risks should be discussed with the patient and the decision to continue therapy should be a joint one between the doctor and patient. The doctor should continue to see the patient at regular intervals and periodic efforts should be made to determine whether the patient can remain symptom free without the use of the benzodiazepine.

It is desirable to treat such patients on an 'as necessary' basis. Most patients find that as their anxiety waxes and wanes, their sleep deteriorates or improves. The patient should be encouraged to reduce the dose or stop treatment for a while when the level of distress falls, but should have a small reserve of tablets available in case of a return of symptoms. Not only does this technique reduce the risk of dependence but many such patients, armed with the confidence of available therapy, will achieve long periods of normal sleep without tablets. When any such sedative, whether it is one of the benzo-diazepines or not, is withdrawn after a period of several months of continuous use it is important for the dose to be reduced gradually, for by so doing the risk of withdrawal effects of any type is reduced.

Withdrawing the Sedative-dependent Patient

Many, though not all, such patients have a 'dependence personality' and there is a distinct risk that in withdrawing them from one substance another will be substituted. It is vital to ensure that the patient has access to one of the safer compounds during the withdrawal stage. For this reason most doctors now consider it desirable to substitute a long-acting benzodiazepine in the early stages if the patient is currently dependent on one of the barbiturates. A large enough dose should be given initially to ensure that the patient does not suffer from any withdrawal effects as a result of the substitution.

It is also important to appreciate that withdrawal can only be accomplished successfully if the doctor is prepared to provide adequate reassurance. The process is an unpleasant one analogous to withdrawal from alcohol, and the patient will need a lot of support. The best results are achieved when the patient is convinced that the doctor will make every effort to ensure that the withdrawal is as comfortable as possible. The doctor should make sure that the difficulties are fully known to the patient and the advantages understood.

Dosage reduction should be undertaken gradually. One useful method is to reduce the dose at weekly intervals by a quarter of the current dose. With a long-acting benzodiazepine this can be accomplished in part by reducing the dose on some days. If at any stage in the reduction severe symptoms appear then the dose should be raised to that which was acceptable. Then, after an interval, the dose should be reduced again, but by smaller steps or at longer intervals. The time for withdrawal in most patients will be about 4–8 weeks. At the same time that the dose is being withdrawn advice should be given to try to ensure that a more regular and normal sleeping pattern is achieved – see page 55.

Withdrawing the Sedative-dependent Patient

Appendix 1

Some Guidelines for better sleep

(1) Rise at a regular time in the morning even after bad nights. This may strengthen the circadian rhythmicity of sleep and wakefulness, and lead to a more regular time of sleep onset.

(2) Sleep adequately but not excessively. Excessively long times in bed may lead to fragmented and shallow sleep.

(3) Take regular exercise during the day. Regular exercise encourages sound sleep, but occasional bouts of exercise do not.

(4) Keep a comfortably cool room. A hot room disturbs sleep, though a cold room does not help to deepen sleep.

(5) Do not go to bed hungry. A light bedtime snack, e.g. a warm milk drink, helps many people to sleep soundly.

(6) Ensure a quiet bedroom. Occasional loud noises disturb sleep even if the subject has no recollection of waking. Soundproofed windows may be helpful.

(7) Avoid caffeine. Coffee and tea lighten sleep, even in people who claim to be unaffected.

(8) Avoid too much alcohol. Alcohol helps people to fall asleep, but the ensuing sleep is fragmented.

(9) Do not try too hard. If sleep does not come easily get up and do something for an hour.

(10) Use sleeping pills only exceptionally. The occasional use of hypnotics is justified to overcome an acute problem, but continued use should be avoided.

Appendix 2

Advice that should be given to patients who receive a sedative

Advice appropriate to the use of all medicines, i.e. taking the dose as advised and reporting back with side-effects, applies to sedatives as to all other medications. Advice specific to sedatives is as follows.

(1) What to avoid.

   (a) Food and drink
   Normal food may be taken, but alcohol should be avoided as sedatives and alcohol increase the effect of each other.
   (b) Other medicines
   Sedatives increase the activity of other depressants of the nervous system. These include tranquillizers, narcotics, other sedatives and hypnotics, antihistamines and antidepressants. If such drugs are prescribed care should be taken with the dose.

(2) Driving and handling dangerous machinery
Because sedatives produce drowsiness, which may persist into the daytime, patients' performance in handling a car may be impaired – particularly at the beginning of treatment. Patients should be cautious when driving. Some doctors have found it useful to start treatment on a Friday night, instructing the patient not to drive his car over the weekend. The patient reports back

on the Monday morning, without using his car if sedation is still experienced. Downward adjustment of dose or continued advice about driving should be considered at that stage.

(3) Depression

A substantial proportion of patients with insomnia are suffering from anxiety together with depression, although the depression may be masked by the anxiety. As the anxiety component is reduced by the administration of the sedative, depression may be exposed. It is important to warn the patient to return if there are any manifestations of depression, or if the disorder does not appear to be improving. In such cases an appropriate antidepressant should be tried.

(4) Pregnancy

There is no evidence that benzodiazepines have an undesirable effect on the fetus, when they are given to the mother during early pregnancy. Nevertheless, on general grounds, like all medicines, they should be discontinued if the patient becomes pregnant, unless there are overriding clinical reasons for their continuation. Since insomnia is not a life-threatening disorder, continuation is rarely necessary.

(5) Stopping sedatives

It is important to appreciate that in most patients sedatives should only be prescribed for a brief period to treat an acute problem. It is useful to advise the patient from the very beginning that a short course is the norm.

Some patients, however, suffer from chronic or recurrent insomnia which may be due to chronic anxiety. In such patients longterm therapy may be necessary. Nevertheless, the patient should always try to reduce the dose or stop it during periods of reduced anxiety with better sleep, only restarting if the anxiety and insomnia become distressing. When therapy has been continued for more than about 3 months, patients should be advised to reduce the dose gradually (over a period of about 4 weeks) to reduce any problems that may occur upon withdrawal.

Appendix 3

Glossary

Automatic behaviour: Brief amnesic periods characterized by rapid onset and recovery, during which the person behaves in an automatic or inappropriate fashion with reduced or no awareness of their behaviour. It is one feature of the narcoleptic syndrome.

Cataplexy and catalepsy: Attacks of loss of muscle tone with paralysis of voluntary movement (cataplexy) or fixed posture, but without loss of muscle tone or paralysis (catalepsy), often triggered by emotion. Cataplexy often occurs as part of the narcoleptic syndrome.

Circadian rhythms: Fluctuations in body functions that occur with a cycle of about 24 hours.

Delta sleep: e.e.g. records which show at least 20% slow waves with a frequency of 0.5–2 Hz or less and amplitudes greater than 75 microvolts. It is the deepest stage of orthodox sleep. It may be divided into stages 3 and 4 but this is now often omitted.

e.e.g.: The initials stand for electroencephalogram. This is a recording of the electrical activity of the brain from electrodes placed on the surface of the head. Characteristic patterns are found during various stages of sleep.

Hypersomnia: Attacks of sleep during the daytime. It differs from narcolepsy in the duration of the sleep attack. In hypersomnia it may persist for hours or even several days. It may occur in association with narcolepsy.

Hypersomnolence: Excessive sleep as judged by normal standards. It is most common in women and cyclical in nature, often coinciding with menstruation.

Hypnagogic and hypnopompic hallucinations: Vivid sensory hallucinations during half-sleep, resembling dreams but occurring just as the subject is falling aleep (hypnagogic) or as the person awakes (hypnopompic).

Hypnic jerks: Sudden jerking movements, often of the legs which most normal people experience occasionally just as they are falling asleep.

Hypnotics: Substances which can lead to sleep. This word is now commonly used synonymously with sedatives, though originally it was applied to substances which were more powerfully sleep inducing. It is now appreciated that this difference is largely dose dependent.

K-complex: Well defined slow negative deflections in the e.e.g. followed by a positive component. They occur in stage 2 sleep.

Kleine–Levin syndrome: A rare disorder in which there is periodic over-eating followed by hypersomnolence (*q.v.*). It occurs mostly in young women and may be related to bulimia.

Narcolepsy, narcoleptic syndrome: Repetitive sudden periods of sleep during the waking hours often under unusual circumstances, e.g. standing or eating. The periods of sleep are usually brief and are sometimes preceded by a sense of fatigue. There is often a family history. True narcolepsy is a highly specific sleep disorder. However, it is more common for the patient to suffer from two or more of the symptoms that form the narcoleptic syndrome or tetrad *viz*: short, but almost irresistible, daytime sleep attacks; cataplexy (*q.v.*), sleep paralysis (*q.v.*), hypnagogic hallucinations (*q.v.*), automatic behaviour (*q.v.*).

Nightmares: Unpleasant frightening dreams that occur during the normal REM stage of sleep. They should be distinguished from night terrors (*q.v.*).

Night terrors: These are terrifying dreams that occur during delta sleep, and hence usually early in the night. They are accompanied by extreme physiological arousal and are rarely remembered subsequently. They should be distinguished from nightmares (*q.v.*).

Nocturnal bruxism: Excessive tooth grinding during sleep which often disturbs the partner more than the sufferer.

Nocturnal enuresis: Voiding of urine while asleep. It is normal in very young infants, but becomes a problem when it persists into childhood.

Nocturnal myoclonus: Highly stereotyped leg twitches which repeat themselves every 20–40 seconds, last from 5 to about 100 minutes and alternate with normal sleep.

NREM sleep: Otherwise referred to as non-REM sleep or orthodox sleep. These are the other stages of sleep that are distinguished from REM sleep. Three stages are usually recognized – stage 1, stage 2 and delta sleep (3 and 4).

Obstructive sleep apnoea: See sleep apnoea syndrome.

Paradoxical sleep: Another name that is given to REM sleep (*q.v.*)

Parasomnias: Abnormal behaviours occurring during sleep. It covers such activities as sleep walking, nocturnal enuresis, bruxism.

Primary insomnia: A rare disorder in which there has been a considerable reduction in sleep from birth. It is presumed that it results from an imbalance in the brain centres that are responsible for the cyclical nature of the normal sleep–wake balance.

Pseudoinsomnia: a term that has been used to define the condition in which there is a complaint of poor sleep but in which observers determine that sleep is normal. It represents too great an expectation of the normal length and quality of sleep, particularly among the elderly.

REM sleep: The initials stand for 'rapid eye movement'. It is a sleep stage which occurs for about 15–25% of the night in normal people, and is associated with the periods of dreaming. It is also known as 'paradoxical' sleep or 'the D-state'.

Sedatives: Substances that are depressant to the central nervous system and lead to relaxation and sleep (see also hypnotics).

Sleep apnoea syndrome: This is also referred to as obstructive sleep apnoea since in many cases there appears to be an obstructive element to it. There is cessation of breathing, sometimes for well over a minute despite vigorous attempts by the respiratory muscles to draw in a breath. It is often associated with snoring, restlessness and drowsiness and automatic behaviour during the day.

Sleep drunkenness: A sense of impaired motor control and unsteadiness which immediately follows awakening.

Sleep paralysis: This resembles cataplexy in that there is loss of muscle tone and paralysis of voluntary movement. It occurs at the commencement or end of sleep and is not triggered by external stimuli.

Sleep reversal: Reversal of the normal sleep–wake cycle, with sleep during the day and restlessness at night. It is usually a symptom of organic brain disease.

Sleep spindles: Bursts of activity in the e.e.g. to 12–14 Hz lasting 0.5–2 seconds. They occur in stage 2 sleep.

Somnambulism (sleep walking): Walking in an automatic fashion while asleep.

Tranquillizer: A substance having a specific depressant effect on the central nervous system such that it produces relaxation and relieves anxiety but does not induce sleep. In practice it is now known that all tranquillizers which are currently available are sedative at higher doses, and the term tranquillo–sedative is a more accurate description.

Index